Feature Films on a Low Budget

1

Feature Films on a Low Budget

John Randall

Focal Press
Boston London

Focal Press is an imprint of Butterworth-Heinemann.

Recognizing the important of preserving what has been written, it is the policy of Butterworth-Heinemann to have the books it publishes printed on acid-free paper, and we exert our best efforts to that end.

Randall, John, 1919-
Feature films on a low budget / by John Randall.
p. cm.
Includes index.
ISBN 0-240-80097-4 (paperback)
1. Motion pictures—Production and direction. I. Title.
PN1995.9.P7R35 1991 90-48113
791.43'023—dc20 CIP

British Library Cataloguing in Publication Data
Randall, John
Feature films on a low budget.
1. Cinema films. Production
I. Title
791.430232

ISBN 0-240-80097-4

Butterworth-Heinemann
80 Montvale Avenue
Stoneham, MA 02180

10 9 8 7 6 5 4 3 2 1

Printed in the United States of America

Dedication

This book is dedicated to all aspiring filmmakers who, for one reason or another, have decided to follow this difficult path to fulfillment, and to my favorite nephew, Hector, whose questions led to its inception.

Special thanks should go to my patient wife, Virginia, who has put up with my manifold eccentricities and idiosyncrasies over the years. To my many friends, both here and abroad, I can only say, thank you for having made my career more fruitful and for the pleasure of having known you.

Just as a priest is wedded to the church, so is a filmmaker united to his art. As it is richly satisfying, it is immensely frustrating in its accomplishment and, because of that, the real intent of this work is to smooth the path a little and make the going easier. May it inspire to perseverance and, in the end, success.

Contents

Preface

Today more people than ever before are studying the art and technique of the motion picture, many of them aspiring to participate in the film industry. It is well known that the doors to the major studios have been virtually closed to newcomers for many years. The personnel who are contracted usually come from other media, such as stage or television. The number that ingress is quite small and the process very selective.

On the other hand, independent production companies offer a much wider gamut of opportunities for people with a basic knowledge of how this type of production is put together. Although the technical procedures are quite similar to major production efforts, the methods utilized on independent films may differ considerably. We will look more closely at the causes and effects of these differences.

As I sit down to write the preface of this book, which is directed to the necessities of low-budget motion picture production, today's newspaper lies on the desk in front of me. An article on the financial page is, in essence, one of the justifications for the existence of this book. It is about the cost of money.

We are speaking of the cost of the money to be invested in a picture, not the cost of production which may vary considerably in different production centers.

Does this high cost of money, coupled with an all pervasive inflation, mean that we should stop making films? Definitely not. It does mean that we must reassess the manner in which films are made, and we are certain that this applies not only to modest productions, but to large-budget films as well, many of which fail to make back their cost at the box office even after reediting, retitling, and determined, persistent publicity campaigns.

However much interest rates may rise or fall, low-budget films are being made and will continue to be made. These films represent a starting point for newcomers and a way of life for many filmmakers.

The four pillars necessary to achieve success are: (1) A good *script*; (2) A well-conceived *breakdown* and *shooting schedule*; (3) A *knowledgeable, versatile, efficient director*; and (4) A *professional crew*.

I would like to point out that this book is not a substitute for the many excellent texts available on cinematic technique, but rather is *supplemental* to them in the sense that it refers to the way techniques can be utilized directly to produce savings in time and money on low-budget productions.

My first experience with low-budget filmmaking occurred many years ago when, as an actor, I was called upon to portray the part of a sadistic Nazi officer of the S.S. The picture was being shot by a Dutch refugee who had had first-hand experience with real Nazis, and he was the writer-director-cameraman. The entire picture was filmed with a Bell and Howell single lens, spring-driven, 35mm Eyemo, using 100-foot rolls of black and white negative. It was low, low budget. Several years later I saw the picture in a theater, dressed up with titles, sound, and music. It wasn't bad at all! I was left with some stills to confound my friends and fond memories of my first movie role. What I did not know was that I had contracted an incurable virus of the most insidious and recurring variety, filmmaking, and worse yet, of the low budget type!

The following years were an interwoven patchwork of film, theater, acting, documentary shooting, television, all in great disorder, and finally the pinnacle, I became a hyphenate, a writer-director. It is hoped that some of this experience can be passed along to the reader of this book. The word *hyphenate* in the film world can usually be equated with experience. One well-known low-budget filmmaker is a producer-writer-director-cameraman-editor who also supervises the distribution of his films! The moral is "Get all the knowledge that you can." You may have to start at the bottom gaining skills as you go along. Just don't *stay* on the bottom.

Some readers may ask, "Doesn't making low-budget films using these methods compromise you ideals?" Definitely not. The Dutch refugee mentioned a moment ago was making a definitive statement regarding his opinion of Nazi brutality. We are not discussing the subservience of content to form, but the better utilization of form to express content.

Many of the statements made in this book are personal opinions and must be regarded as such. They are not dogma. Your reality may be different from mine. What I do expound is the message that method rather than anarchy must be applied to making a picture. It is my hope that professionals who are locked into a single activity of the industry will also find something of use to them here, as well as those who are beginning in the field, because it is a compendium of the exact steps followed in the making of commercial low-budget pictures. For the latter, remember that filmmaking is one of the most all-encompassing, extensive, and comprehensive activities invented by humans for the enjoyment of other people, and all roads lead to the finished work. Good luck, and may you enjoy your virus as much as I have mine.

John Randall

Acknowledgments

When one writes a book on a complex subject such as low-budget motion picture production, one owes a large debt to many people. It can truthfully be stated that over the years I have learned what to do and, sometimes, what not to do from directors, cameramen, technicians, art directors, actors, editors, and composers with whom I have had the good fortune to work. The public, with their favorable and sometimes unfavorable responses, has also taught me much about this phenomenon of the entertainment world, the motion picture.

The comforting words of friends have helped when the going was tough. The wise words of some directors were of importance and value. As a director myself, I am grateful for the performances given by players and technicians who have contributed to my work.

Opposition is valuable, too, and should serve as a stimulus causing us to learn more while striving to overcome it. When we do this — when we oppose, overcome, and triumph against any and all odds — the victory is even sweeter. And remember your critics; they are valuable, too. Their biting comments almost always contain some grain of truth that can serve as a stimulant, producing an intellectual "antibody," so to speak, which provokes an interior reaction to show them that it can be done in spite of all words to the contrary.

1

Writing the Screenplay

In the beginning there was the word. The screenplay is a blueprint of a motion picture constructed with words intended to be read with creative imagination. If extended with thoughts, nuances, and lengthy descriptions, it would be hundreds of pages long, confusing rather than enlightening the users. A properly written script contributes, in a sparse manner, all the necessary information to carry out the elaboration of a picture by the producer, director, cameraman, editor, and any other personnel related to it.

All of us begin our relationship with the cinema in front of the screen as a member of the public and gradually penetrate toward the rear until we become the illusionist dominating the mechanics of the effect we wish to produce. Slowly we are converted from the listener to the storyteller. This is when we must become aware of the two faces of the script.

The screenplay tells the story in cinematic terms rather than in the play form used in the theater, or in the novel form, and its elaboration should conform to certain standards to facilitate production. From its very inception, it is possible to predetermine the approximate cost and shooting schedule of a picture utilizing the screenplay. This places the screenplay foremost when considering a low-budget production.

Why discuss the screenplay before going into the main course of low-budget production? Because, quite literally, the script is the most essential element utilized in the elaboration of motion pictures. From the producer to the editor, all view it from different angles and sometimes draw completely unique conclusions from it.

In the area of screenplay writing, there exists a problem of semantics, a confusion between the meaning of two quite simple words: what and how. The "what" of a script is the story and that is as varied as the imagination. The "how" to get it up there remains the same. This is the great dichotomy of motion picture writing, for all screenplays have two distinctly different lives, one related to what is to be produced and the other to how it is to be produced.

A completely detailed script is an absolute necessity. With it in hand, a breakdown, shooting schedule, and preliminary budget can easily be prepared. It's the equivalent of a chart for the navigator: Without one, you can't tell in which direction to go. Once you have a solid base on which to work, you can construct an edifice of amazing proportions.

Before we go into detail concerning the practical elaboration of these blueprints, let's take into account some general principles. First, the story line. If this presents a story for an epic film related to the fall of Troy, we are talking of an historical costume picture that will, consequently, cost more than a contemporary drama. This does not mean that you cannot or should not attempt the historical picture, but you should consider the implications of doing so.

Another aspect relative to the cost concerns the numbers of geographical locations. One day my phone rang. It was a friend who was going to produce and direct a children's picture about a boy, a girl, and a horse. He invited me to participate as executive producer. I said, "Tell me more." He began reading a detailed synopsis while I, with the phone in one hand and pencil in the other, took notes. When he had finished, I had enumerated fourteen widely different geographical locations with the consequential transportation, housing, and catering problems.

"What is your tentative budget?" I asked. He told me. I think I held my breath for a moment and then responded, "It's going to be difficult, if not impossible, to make it for that sum."

"We have a completed screenplay and over 600 story boards made up for the picture," he answered. "It will work!"

I had no choice but to decline. As he was a good friend and still is, I explained why I couldn't take part in the venture — because I thought the final cost would be much more than they had calculated due to the long distances involved and the high cost of transporting equipment, personnel, actors, and, needless to say, the horse, which would definitely put the expenses higher than was originally estimated. He was not deterred, however, and proceeded with the project. The producer put up the money and the project began to roll. The two teenage actors were terrific, as was the horse, but about halfway through the schedule the production came to a disconcerting halt. The money had run out.

The producer's first and last venture into motion pictures had turned into a disaster, for he had no more available cash and none was forthcoming from other sources. Teenagers grow fast and there was no time to wait. The very unhappy producer finished the project with some worthless cans of film, and the director with a tarnished reputation. Reality, distasteful as it may be, must be faced from the very first page.

Do not start a picture without the money to completely finish the principal photography.

Does this mean that we must forego using location filming or a cast of thousands? Not in the least, but it does mean that a picture requiring these things must be planned in a special way. The classic example is the football stadium with thousands cheering, cut to a tight shot of a section of the bleachers where the hero feeds the heroine popcorn, surrounded by cheering neighbors. Or the race track; cut to rail birds winning or losing. The boxing arena in long shot; cut to low-angle shot of our hero battling it out with the villain of the piece. Pick-up shots of exteriors lend realism. All of these examples and many more, including rock concerts, have been and will be used in low-, medium-, and even big-budget productions. It's not *what* you do, but *how* you do it that determines the cost and the end results, just as in a lot of other things.

So, if you are buying a screenplay or writing one, count your characters, count your locations, then count your dollars. There's a definite relationship, so don't expect your low-budget picture to look like the production of *The Fall of Rome*. This is not to say that a personal definitive statement of that subject could not be made on a small budget, but it would have to be written taking into consideration the financial limitations inherent in the project.

Many popular novels have served as a basis for successful pictures. Adapters are continually working on them; however, the rights to a novel can be very costly unless the work of a beginning novelist can be obtained prior to publication. Otherwise, delve into the past. Most of the great novels are in public domain. A well-known company produced practically all the works of Edgar Allan Poe, including his poems, sometimes using only his titles.

The past is a veritable gold mine for the independent producer, but he or she must be absolutely sure that these works are in public domain. Beware of translations from other languages as they are usually protected by copyright. If necessary, go to the original — just look before you leap.

Another important factor from the budget point of view is the number of sets and their utilization. Both the writer and the director should make the most of the materials they have available.

When we look at films made by Lumiere and others during the infancy of the industry, we see that they placed the camera in one position and cranked it through the entire action. With Melies the medium began to change, and D.W. Griffith brought film almost to its maturity. A few years ago a Japanese director became an artistic success following the one-camera method. While this method makes for a cheap picture, it usually results in an epitome of boredom.

Whether your financial status dictates a low budget or not, the public has reached a certain stage of sophistication that demands, and must have, reasonably competent workmanship, from the script to the final print.

I recently saw an early film by an Italian director (now deceased) of cult fame. It was painful to watch. The jump cuts, the bad framing and general technical poverty considered "arty" overwhelmed whatever modest social content the picture may have contained.

Moving camera, close-ups, action, and utilization of technical resources are rightfully expected in today's films, and these should be written in from the very beginning.

For example: You have eliminated the crane shot in the script because of high rental cost, but suddenly you see a neon sign business close to where you are shooting and they have available a giant hydraulic crane with a basket for installing and working on signs. You may rent it cheaper than the special crane used in the studio and you can put the shot back in the script on the spot. Be resourceful and use anything that can help save money and enhance the picture. The same applies to airplanes, helicopters, boats, etc. Write them in, try and get them, but don't make the shot contingent on their use.

Some authors state categorically that a completely detailed script is not necessary and that only a cursory description of the content of the scene is sufficient, allowing the director leeway to interpret as he or she sees fit. This leads to a picture over schedule, over budget and, sometimes, to general disaster. Every shot should be written into the script. This is not to say that the director doesn't have options. As director, you can dictate the complete shot and how you wish it made, but the fundamental information contained in the script is necessary for the elaboration of breakdowns, budgets, shooting schedules, etc. The entire sequence must be written down on paper so the producer, director, players, cameraman, editor, script supervisor, and accountant know more or less where they stand. This is preplanning versus postmortems!

Improvisation falls into the same category. It leads to madness, self-gratification, and trouble. Just because a well-known director on one occasion allowed his actors to improvise all their lines and managed somehow to finish the picture doesn't mean he saved money, made money, finished on schedule, or even made a good picture. Not that changes are not or should not be made in dialogue. They are made all the time, because a line does not read right for a given actor, or the tempo is not correct, or its accent isn't acceptable — any number of reasons.

Another aspect is that the player is working from a definitive base provided by the writer, but in accord with the player's own general concept of the play and your interpretation of the same as the director. With improvisation, the smartest and most vocal actor is, in essence, writing the dialogue for you, which may or may not be in agreement with the writer's original concept of the scene. The most spontaneous lines are usually thought out and written down well in advance.

At this time there is another very important factor to consider. Giving only the general idea of a scene is the lazy writer's approach to a quite concrete problem. The script must be completely finished, with no holes in it. It doesn't matter if a certain director wrote the next day's scenes the night before the shooting. Sometimes this may have to be done as a desperate last resort, but it usually doesn't work. In practice you will find one change begets two more changes, which beget four more, with the result that the story is soon in a shambles and, in the end, is an entirely different picture. Solve all your little writing problems at the proper time, before they become roaring lions on the set.

Now for the nuts and bolts of screenwriting. For a person in a strange country, any map is better than no map at all. The system I put forth works for me. It's simple, easy to follow, and it does lead to a finished screenplay. Whether it achieves its goal depends upon the original story, your talent, and many other interrelated factors. Your ability to visualize the scene as a camera would see it is quite important. You must know in your mind what is encompassed in a long shot, a medium shot, and a close-up. Your dramatic sense must tell you when to cut any given shot — but hold it, we are putting the cart before the horse.

First, write out a detailed synopsis of your story. While being concise and to the point, your synopsis should contain all the pertinent dramatic elements, presenting the characters involved in the order of their appearance, and ending in a good, satisfying climax. While working on the synopsis, you'll have to make some very basic decisions. As mentioned before, you must decide on the number of locations and characters, who they are, and what part they play in the story.

To help you construct the play, make a list of these people, giving them names, characteristics, and peculiarities. Give as much detail as you possibly can under each name. Remember these characters are the children that you are going to bring to life, and much closer to your heart's desire than your own children. Both their physical and psychological aspects must be taken into account at this time, including a resumé of their past experiences. Although this may never appear in your finished work, it will serve to form your concept as to how they should or would behave under any given set of circumstances. You must know these persons very well indeed. To prepare information on the principal characters, study *Figure 1* for guidelines. It will help you throughout the process of writing to visualize the actors in any scene.

When you have finished the synopsis and the list of characters, you can start to work.

The following steps will determine the architecture of your picture. Just as a well-proportioned building has to be consciously designed, so must a screenplay be laid out. Just as most buildings don't look alike, so most screenplays won't be the same. The fact that a blueprint is used does not necessarily mean that any two

Know Your Characters

1. Name, Age, Race
2. Height, Weight
3. Color of Hair and Eyes
4. Physical Description
5. Education, Occupation
6. Social Class and Economic Status
7. Married, Single, Divorced
8. Religion
9. Father, Mother, Brother, Sister, Orphan
10. Wife, Husband, Girl or Boyfriend, Lover
11. Honest/Dishonest, Likable/Unlikable
12. Handicaps, Fears
13. Abilities, Skills, Intelligence
14. Extroverted/Introverted, Optimist/Pessimist
15. Relation to Other Characters
16. Protagonist/Antagonist, Hero/Villain
17. Goals, Triumphs, Failures

▶ ***Figure 1*** *Before plotting the screenplay, you should establish the principal characters as completely as possible. Use file cards for annotations.*

STORY OUTLINE FOR DEVELOPMENT

1. Kitty dances in Nick's night club. They are forced to pick up contraband using his yacht.
2. The Coast Guard pursues Nick's cabin cruiser. Nick and Kitty arrive at an island.
3. Bancroft tells Nick that he has no gas. Nick and Kitty are invited to Bancroft's house.
4. Bancroft is a writer. Nick and Kitty have dinner at his house.
5. Nick listens to the Coast Guard on the radio. Kitty changes her clothes.
6. Bancroft, Nick, and Kitty play dominoes.
7. Nick and Kitty make love in the cabin cruiser.
8. Nick hides the contraband in the dunes. Bancroft spies on him.
9. Kitty and Bancroft walk on the dunes. Bancroft invites Kitty and Nick scuba diving. Nick refuses.
10. Bancroft and Kitty visit the wreck of a sunken ship. Mishap of Kitty.
11. Kitty tempts Bancroft in an outdoor shower.
12. Nick walks along the shore looking out for the Coast Guard, Kitty returns to the cabin cruiser.
13. Bancroft digs up the contraband late at night.
14. Bancroft invites Nick to play dominoes.
15. Kitty, feeling slighted, brings a record player and dances for them.
16. On return to the yacht Nick beats Kitty brutally.
17. Later Kitty begs forgiveness but Nick rejects her.
18. Kitty goes for a swim nude. Bancroft encounters her on the beach.
19. Kitty and Bancroft start to make love. Nick sees them.
20. Nick and Bancroft fight. Nick is wounded. Nick dies in Bancroft's house.
21. Bancroft and Kitty bury Nick. Kitty decides to stay on the cabin cruiser.
22. A tropical storm blows up. Kitty flees in fright from the yacht. Bancroft searches for her.
23. Bancroft makes love to Kitty as she fantasizes with the memory of Nick.
24. Kitty realizes that she is repelled by Bancroft.
25. A Coast Guard planes flies over the island looking for Nick.
26. The Coast Guard arrives to the island to arrest Nick.
27. Bancroft delivers the contraband to the Coast Guard.
28. Kitty reproaches Bancroft and they have a bitter quarrel.
29. Bancroft follows Kitty to the dunes begging her to stay. After mutual recriminations they part in anger.
30. Kitty is seen leaving with the Coast Guard and Bancroft returns to his typewriter.

THE END

screenplays will be the same. The use of the blueprint is a discipline that produces and controls the desired result.

To start this underlying structure, take a sheet of paper and put down numbers one to 30 on the left side of the page.

Write the title of the first sequence after number one and the last under number 30. In addition to this, put the sequence number with its heading at the top of a filing card, and jot down on this all the pertinent notes, additions, details, bits of business, and dialogue, until sufficient detail is built up to begin writing the first rough draft or treatment. It is a good idea to concentrate a maximum amount of usable material on those 30 cards. Start spacing your dramatic points after the intervening numbers. Remember that a 90-minute picture must have at least six or seven dramatic, gripping scenes, and for 120 minutes of screen-time the script should have at least eight of these, each point contributing to increased interest, tension, or suspense leading up to the grand finale of the picture.

After you have been writing for some time, you will notice suddenly that there are a lot of blank numbers on your list. You have arrived now at the time where the story, which may have begun as an anecdote, must be fleshed out to convert it into a full-fledged script. Unfortunately there is no easy way of doing this. You may wake up at 3 A.M. with an idea, or you will unexpectedly think of it as you try to move through the jammed-up traffic on the freeway. Your friends will notice a glazed look in your eyes at odd moments, and if they know you well, will be aware that you are thinking about "the story." The bits and parts, the additional episodes, the details, will gradually fall into place until everything jells into a complete entity.

By using the numbered sequence system, all the part of the jigsaw puzzle can quickly be identified and put together, making the actual writing easier. With this method you will have (1) eliminated all the major potholes; (2) taken note of all the felicitous phrases and bits of business to be used; and (3) placed all the pertinent developments in the correct order.

For low-budget filmmakers, the terms *screenplay* and *shooting script* are synonymous, which means that the screenplay should be written as shooting script from the very beginning. The page with the numbered sequences and the filing cards are in front of you. It's time to begin writing in earnest. Put down number ONE with the pertinent information after it, i.e., EXTERIOR or INTERIOR, then the location, DINING ROOM, for example, followed by DAY or NIGHT as the case may be. Under this, write a brief paragraph describing the action and camera movements, including briefly any other pertinent information. If there is dialogue, identify the speaker with capital letters, placing the dialogue itself under the name. Continue through all of the sequences until the script is completed. When you are through with page 120 or so, you will think you've finished but you haven't. Take a short break and get away from the script. The change of pace will give you perspective and you can come back to the work with a fresh approach.

You will notice now that many details are still missing from the script. Do not be alarmed, for this is normal with all rough drafts. Start over from the very beginning, studying each setup, scribbling notes between the lines and polishing it for the rewrite, adding takes or removing this or that, making changes that will thoroughly

SHOOTING SCRIPT

219 EXTERIOR — TERRAZA — EVENING.

Close-up of hands mixing dominoes. Empty ashtrays, pad, and pencil. Nick mixes.

219A EXTERIOR — TERRAZA — EVENING.

Long shot of the terraza. Bancroft, Kitty, and Nick are playing.

219B EXTERIOR — TERRAZA — EVENING.

Extreme close-up of dominoes on table. The camera dollies back to reveal full ashtrays, empty glasses, pad full of numbers. As the dolly continues back we see Bancroft, Kitty, and Nick playing. The game is almost finished. Bancroft places a piece and Kitty follows.

NICK

That doesn't go there.

Kitty picks up the domino and puts down another.

KITTY

Shucks — I must be stupid. I can't play this game.

BANCROFT

Bancroft puts down the last domino.

This one is finished (turns to Kitty).
It was my game anyway.

NICK

(Looking at Kitty) So, it's a kid's game?

Nick dumps down his hand and begins to count. Kitty counts her hand too. Nick takes a pencil and put down numbers. Kitty gets up and walks to the rail. Bancroft watches her and rises also. Nick slams down the pencil and get to his feet.

220 EXTERIOR — TERRAZA — LATE EVENING.

NICK

Let's get going, Kitty.

BANCROFT

(Turning to Kitty) Want a drink before you go?

upset your first sequence of numbers. Do not despair, for it won't be the first time — or the last — you will have to change the numbers. After the shooting script is finished, just write the changes as 312 A, B, C, or D, as the case may be, without renumbering the whole script.

Warning: Make all the changes before you send the original out for copies and most certainly before you make the breakdown or shooting schedule. If you don't "freeze" or stop making modifications in the script at this stage, the subsequent changes will cause anguish later on.

After the second rewrite, with most of the typing and spelling errors corrected, we have before us a completed 120-page screenplay with approximately 500 to 800 numbered scenes: the pages are fairly clean and legible and we are ready to put it together. Now you must decide on how many copies are necessary and who is going to have them. Let us suppose that you can make this decision. If not, stop worrying and hand it to the producer. It so happens that everybody, including the person who serves lukewarm coffee to the cast and crew, would like a copy but, remember, it's not Christmas yet. The budget will only permit basic necessities. Fifteen copies are usually enough. Remember to send one to the Copyright Office, Library of Congress, Washington, D.C. 20559, along with the $10 fee and a completed Form PA. Within a few weeks you'll receive your stamped certificate of copyright registration. This will certify that the screenplay is your original in case someone tries to sue you and also gives you the basis for a lawsuit in case the copyright is stolen outright or infringed upon in any way. Also, register your script with the Screen Writers Guild.

Outright plagiarism is rare but not unknown. Unscrupulous competitors will steal your ingenious tidbits that differentiate your work from that of others and use them in their picture. Usually there is not recourse against this piecemeal theft.

On this note, let me give you some very pertinent advice. Don't show your script around and don't give it to other people to read and comment on or about. Amateur criticism is more devastating than professional, it that's possible. The only person who has the right to suggest changes is the producer, who is paying for the script and is going to risk money to make a picture based on it. If you are sure of yourself and your creation, stick up for it, but not to the point of obtuseness. One other note about producers: Offer to make small changes, but do not, under any condition, make all the rewrites the producer may think necessary or you'll spend the rest of your life rewriting the same story and starve to death in the process. The producer should be satisfied with a good, workable shooting script. Remember, it is not a literary masterpiece but the "organized presentation of a story in the correct cinematic form to be utilized in the elaboration of a motion picture." All else is beside the point.

Take the typewritten pages to a copy establishment that has a collating machine; choose a good sturdy cover and order the copies, being sure to use General Binding Corporation's (GBC) flat plastic hinge. GBC's line of rectangular holes allows you to lay the finished script flat or double it back on itself for easy use. The blank back side of the previous page, which faces the page you are using, is ideal for notes, corrections, additions, drawings, etc. The script supervisor will be happy because extensive notes can be written on it without having to twist the script into a roll. The ordinary stapled and bound script is quite difficult to handle. If you need an extra page or two for small roles, just fold the script to the pertinent pages and make the number of copies required. If this criteria is applied to all items, it's possible to save considerable sums in the long run.

At this point you will find yourself faced with a tremendous decision: "How long is this picture going to be when it is finished?" or, should we say, "How long is this picture intended to be?" A finished shooting script should contain enough material so that each page represents approximately one minute of screen time. And you must ask yourself literally, "Where is this picture going to be exhibited?" It has never been easy to answer this question, as producers, writers, and directors have found out; but with the advent of film markets and television there are more opportunities than ever before.

Where do low-budget films really go? Today there are roughly three domestic market areas to be considered: drive-ins, small theater chains and television, and video cassette sales. Forget simultaneous openings in 1200 houses nationwide; that's for the majors. For the moment, stick to reality and let the fantasy wait.

The completed registered script is the first major step in motion picture production — that is why it occupies this first chapter — but there is another reason, which is, what kinds of films do exhibitors prefer? To start with, nothing less than 90 minutes. Some of the theaters still play double bills, so if the film is over 120 minutes, you present a problem to the exhibitor, who may have trouble showing two films without finishing the show very late at night. Also TV time slots are just right for 90- and 120-minute pictures. With a few judicious cuts made for the tube viewers (really to make space for the commercials), everyone is content.

Another important preliminary decision that must be made when writing the script concerns the Motion Picture Association of America (MPAA) rating to be desired. "Why at the beginning?" you may ask. "Isn't that a little premature?" Yes, this is the time and place. Unless you have a very, very good family picture with a G-rating to compete with Disney's latest Buena Vista production — forget it.

If you jazz it up a little bit and end up with a PG, you are worse off than before. PG-13 is a dubious compromise, best ignored for low-budget pictures.

Where does this leave us? There is only one choice left, and that is R. The R-rated pictures are accepted all over the world, from Argentina to Zanzibar. If, as an independent, your picture receives a good solid R-rating, it means that you are in business — or should be. Write your screenplay with this in mind and make everyone happy, including yourself.

Features with musical numbers should be spaced out correctly like any other work. It is always difficult to hold a good, solid dramatic line with interspersed songs and dances. But proof that it can be done quite well resides in one of the finest examples of this genre, *Cabaret*. It must be remembered that several relatively low-budget producers, like Joe Pasternak, were very successful with this formula during the 1930s and 1940s.

If you want to write a low-budget musical, go ahead and do it. The money-saving secret in this type of picture consists in having every little detail worked out in advance and thoroughly rehearsed until ready, even before the set is rented or constructed and the crew hired. Some directors, like George Sydney, worked out a complete storyboard continuity before beginning the picture. Improvisation on the set of a musical, with time passing and money flowing out inexorably, can be disastrous for the budget. Fortunately screen time can accumulate quite rapidly in the production of a well- prepared musical. In recent years several rock musical productions have been shot using 16mm cameras, thus reducing the cost of the picture.

Action pictures present a serious problem for the writer of low-budget independent features. The public has seen so many really good, thrilling, action pieces with the destruction of planes, helicopters, buildings, cars, and diverse property, both in models and life-size, beginning with the top register of James Bond and finishing with the hazardous material seen on TV, that it is hard to come up with something similar on a low budget.

In many pictures the chase is the *piece de resistance*. If you have one in your story, make it as good as you can within your limitations. (We are sure that with ingenuity you can do a lot.) A foot-chase is always good for low-budget work. I shudder when I remember having seen one, some years ago, in a very low-budget feature, where all the action was shot from one camera position. Use your imagination. It's just as easy to do it well as not, and it will be easier to film when well structured. If it is not completely written down on paper, anything can happen. "A chases B through Central Park" is not enough. When all the setups are specified, the director can add or subtract from them as appropriate, doing a much better job in a shorter period of time than having to devise it all on the spot. In the latter case the director is doing the writer's work as well. Make it easier and the result will be a much better picture.

There is another reason why the script occupies the first chapter of this book: a script has value. It is an asset of the production company with a value of $5,000 to $500,000, even unproduced. Many producers have several scripts in different stages of preparation. It also follows that if a writer has several good, well-prepared properties to offer to a producer the writer's value will be enhanced and there is a better chance of having the work accepted.

In the foregoing pages I have examined a method of setting down on paper sufficient dramatic literary material to make up a shooting script of a certain length. The format has been presented; now I must speak of cinematic grammar. A writer conversant with this grammar can, by stipulating the shots, angles, camera movement, and action of the players within the scene and the dialogue, effectively dictate to a great extent what will be seen on the screen. When the script is written with ample pertinent detail, the writer can project his or her creative effort to the maximum at the same time, providing the producer material to construct an effective budget and the director a solid base upon which to exercise his or her art and craft.

The first part of this grammar consists of the various shots and their usage as they, in essence, cover the area that we will see.

The *long shot* or *extra long shot* are used to establish the site of an action. Although a sequence may begin with another shot, at some time it is usually necessary to reestablish the scene. If an establishing shot is not used, viewer confusion may result, unless some other form of establishment is present, such as a peculiar sound relevant to the local or relative dialogue. As pictures are a visual medium, it is always preferable to show something rather than talk about it. It is convenient to notice that the presence of a player is not obligatory in these shots, although it may help to relate the shot to the sequence.

Even though the long shot has not been a favorite in television plays, it should not be despised. The long shot reached its apotheosis in Western Cinemascope dramas covering the entire panorama of a giant screen with all the outdoors.

Probably the most useful and the most overworked of all the camera positions is the *medium shot*. Practically all of the action is carried forward with it. As it is so important, it occupies a great percentage of screen time and, for this reason, requires special attention. A long series of static medium shots ultimately ends in boring the viewer. In order to avoid this and enhance the dramatic content of the scene, writers and directors over the years have combined this shot with both movement of the players and the cameras. Technicians have used scenes with great depth of focus and scenes with very selective focus. Incredible fluidity can be introduced into the medium shot, with hardly any additional expense, by the use of a dolly and zoom lens. Just because the medium shot is the workhorse doesn't mean it can't have variety and stimulate viewer participation in the action.

Just as television, due to its reduced screen size, has robbed the long shot of its dramatic impact, live theater never has provided the forceful impact of the *close-up*. This great cinematic invention separates this medium from the others, notwithstanding the terrible abuse and misuse of the shot in TV dramas over the last few years. When placed correctly, the close-up coincides with the peak of dramatic content in a scene, reinforcing its strength. When misused it has no meaning and contributes nothing to the picture. This shot calls maximum attention to itself as well as to the action.

An insert is used sparingly when some very important piece of information is imparted. Letters and written notes should be avoided since viewers in other countries may not understand the content. If the information is vital to the story, it should be presented in some other form.

Closely related to the close-up are the *two-shots* and *over-the-shoulder shots,* also known as *reverses.* These shots, somewhere between the medium shot and the close-up, lend themselves to a great variety of treatments. As with the medium shot, they can be combined with judicious player and camera movement and, although a little more restricted, convey more dramatic information without becoming the great cymbal crash of the close-up.

Walk-in and *walk-away shots* are also quite necessary, allowing characters to become more or less important to the scene. This shot is useful to end a sequence, just as a head-on shot useful to bring someone or something into the foreground. The two shots are quite effective and lend themselves well to making story points.

The *point of view* or *POV shot* is important but, like all other strong medicine, it should be used only when needed. It brings the viewer right into the situation. However, if this person is identifying with another player, the effect won't work, so its application must be judicious. An entire picture was once made from POV, but was not completely successful because the audience was desperately desirous of seeing the protagonist and, when that was impossible, felt frustrated.

Angles in cinematography are intimately related to perspective in painting. They have a profound psychological effect on the viewer. Everyone unconsciously accepts the eye-level viewpoint as normal. In this position the camera does not participate in any change or coloring of the scene, but simply presents it to us. The majority of the shots are usually taken at zero-angle viewpoint. This does not mean that the scene may or may not have dramatic content; it simply means the camera angle has not intervened to enhance, detract, or in some way change this value.

When a *low angle* is used, the viewer is placed automatically in an inferior position and, without being consciously aware of it, is being regressed to a childhood stature where adults command a superior and sometimes threatening position. In this position the viewer feels insecure and menaced in the presence of the villain and adopts an attitude of fear, the intensity of which varies according to the degree in which all the cinematic terms are applied to obtain this effect.

The use of a *high angle* accentuates the superiority of the viewer and conveys a sense of loneliness, subservience, or debility on the part of the player being viewed. This is a difficult angle to accept and must be used sparingly so that it will be effective when used.

The *oblique angle* tends, just like all the other angles, to change perspective. In this case the importance of the enlarged foreground image is much greater than the background one, which is smaller. This change of image size can be used to greatly augment dramatic impact. Its use, of course, must be moderated with discretion like any other technique.

Odd or *unusual angles* are sometimes very useful, but discretion again must be exercised, for when an angle draws attention to itself it is not serving the purpose of the picture as a whole. Too many distractions of this type cause viewers to return to reality, having lost their immersion in the story.

When the camera became free to move, the motion picture came of age. Probably no other change — sound, color, or screen dimension — affected the motion picture as much as the mobile camera. It fulfilled an age-old desire to view an event from all angles, to move, to circle, to run, to fly!

The basic and most-used camera movement is the *pan*. Its simplicity makes it a common integer in an enormous variety of complex shots. It is only when it is performed incorrectly by the camera operator or misused by the director that it suddenly becomes apparent. The "garden hose" pan, going back and forth, and the pan that crosses groups of people with wide open spaces between them come to mind, along with pans that fail to move correctly with the action. The pan should be initiated a fraction after the movement starts and end just before the movement ends. A good pan tells much more about the scene; but don't overdo it, too many pans are worse than none.

Along with the pans there is the *tilt up* and the *tilt down*, both very useful movements, for when combined with a pan they allow us to follow the idea behind the shot, even if there is no movement within it. A *tilt up* from floor level to a person looking down is quite dramatic. With the use of light weight cameras and fluid heads on the tripod, pans and tilts are much more easily carried out than with heavy cameras and gear heads that have been used before.

Two other important movements are the *dolly-in* and *dolly-out*. They convey a sense of active participation because you can follow closely a player as the camera moves in on him to better display his reaction. When the mechanical four-wheeled dolly is used, the shot is usually time-consuming to set up, since the grips who push the vehicle must hit the chalk marks exactly. That takes practice, not to mention the time consumed in laying the tracks. Much to the horror of the purists, we recommend without reserve the use of the zoom lens for *dolly-in* and *dolly-out* shots. In spite of the perspective differences, which are really small if the shot is a tight one, no one can perceive the difference between one or the other. More on the zoom lens will appear in a later chapter.

The *tracking shot*, in which the camera moves along with the players, gives the viewer a sense of participation in the action, in essence accompanying the players as they proceed through the scene. Without separation from them, the viewer is privy to their conversations and actions, producing a certain intimacy and identification with the players. This is a quite useful shot, but remember that it is difficult to cut into a tracking shot and the shot should always be used in its entirety.

The *crane shot* has been mentioned before, but bears mentioning again. If a crane is available, by all means, use it. Of course, if you have to rent a crane with its driver-operator and take it to a location hundreds of miles distant for one shot, give it the axe and find another way to do it. A shot from a large crane outdoors is usually spectacular and can greatly enhance a picture. A small crane indoors makes staircase shots a dream and can prove very useful if you have one.

The *cuts*, which normally would be considered the province of the editor or the director, are very important to the screenwriter. These are direct cuts, jump cuts, cutaways, parallel, subliminal, and intercuts (see the glossary for explanation of these terms). The *direct cut* is usually not written into the script; it is taken for granted.

Cutaways, however, should be written into the screenplay if possible. They are extremely important and should be specified whenever possible. In the first place, they give the drama an added impact, the reaction of someone present or cognizant of the action, and, second, they help the editor. If it's impossible to cut on the action or otherwise go from one take to another without a disastrous *jump cut*, the editor can always use a cutaway to cover up the lapse. These things do happen, sometimes more frequently than we would like to believe. Write the cutaway into your script and you will never regret it.

Parallel cutting should be written into the script, even though the director will shoot both scenes out of sequence with the correctly numbered slates according to the breakdown. The editor will then provide the correct intercutting following the indications in the script.

Subliminal cuts or the *intercutting* of similar material from past scenes into a later scene should be written into the script, but the timing of these cuts must be decided by the director and the editor in front of the moviola.

The screenwriter should be aware of the use of flashbacks and their presentation, which must be clear and precise from the writing stage on. The treatment of time in a script is important, and time lapses are necessary and should be so indicated. However in the telegraphic style of today, many of these time lapses have been eliminated; consequently pictures move along at a faster pace than before, although sometimes it is detrimental to the mood of the scene. A fine balance is required for a good quality screenplay.

Montages are quite useful in low-budget film production where at times it is necessary to present a multiplicity of informative shots in a short time. Don't get carried away with esoteric theories — just be practical and write down what you want, within reason, and the director will get it for the editor to *slice up*. Remember when you write a script, that motion picture time is only remotely related to real time, even in *cinema verité*.

Several good books have been published on screenplay writing. Read them with a critical eye, picking out the parts that are useful for your needs. Above all, be selective; fewer characters and locations, used with greater forethought regarding the dramatic values of each, will lead to a better low-budget picture.

We will finish this chapter with some "optical" explanations. There are fade-ins, fade-outs, dissolves, cross fades, the ever-present freeze frames, wipes, flips, out-of-focus, etc.

Some opticals are old-fashioned now and others so obvious that they can't even be used in commercials. Most opticals are costly and complicated. They are used in pictures every day, but low-budget film makers try to avoid them for those reasons. Be sure to study them so you will know what they are and how they are used.

SUMMARY

- Limit the number of locations.
- Limit the number of characters.
- Write a detailed synopsis.
- Make a dossier on principal characters.
- Lay out a list of 30 numbers as milestones.
- Create additional material or subplots if necessary.
- Prepare cards for each sequence.
- Detail all the shots.
- Eliminate holes in the story.
- Renumber shots and number pages.
- Copy, collate, and bind the script.
- Register the screenplay.

2

The Breakdown and the Shooting Schedule

If it has been your good fortune to work on a Hollywood production, you will have noticed a person rushing about with a long board filled with multicolored vertical strips of cardboard, inscribed with mysterious numbers opposite a list of places on the left and a list of names across the top. In England a person may be seen around the office or on the set of a production with one or two square tablets that look like badly constructed crossword puzzles. Every film production anywhere in the world has its equivalent. The board and the tablets being displayed are the shooting schedule and the breakdown, combined or separate as the case may be, and represent the order in which the picture will be shot.

Why do the director and producer defer to this person? To answer this question you must first consider the script. All the dramatic incidents, all the locations, and all the players are set down in it. The script is a complete document in chronological order, but it tells us little about how the final project will be constructed most efficiently: the order in which the work should be done, what will be used to do that work, how long it will take, and, last and most important for this text, how much the picture will cost. The answers to all these extremely important questions are embedded in the script, but are not visible to the naked eye, so to speak, and must be extracted and interpreted.

For example, after a cursory reading you note that there are many scenes in the living room set. However, after making a breakdown of the script, you find there are 168 setups, some at the beginning, some in the middle, and some at the end of the picture, with corresponding wardrobe and makeup changes, which, according to deductions based on previous experience, will take six working days to shoot and will require the presence of three players.

It's virtually impossible, if not impractical, to shoot these scenes in the original order as set down in the script. It is, of course, proper and correct to shoot each sequence within this large group in their correct order to maintain a unity of dramatic values.

When all the information is extracted in a breakdown of the script it will be possible to do the following:

1. Make up a location list.
2. Construct a cross plot.
3. Make up wardrobe lists.
4. Make up a prop list.
5. Construct a shooting schedule.
6. Construct a definitive detailed budget for the total cost of the production.

Now consider each of these documents separately in depth.

The production office usually does this vitally necessary paperwork but many times, especially on low-budget productions, the assistant director or even the director may have to do it, although the budgeting part is usually carried out by the assistant or the executive producer. For a low-budget production it is absolutely indispensable to carry out all this work as completely and efficiently as possible.

The first part, the location list, is deceptively easy; the hard part comes later. Starting with page one of the script, every new location is noted on a list as per the sample page from the picture *Deadly Reef.* When finished, as in this case, it will be seen that there are 20 different locations. After finishing the breakdown, the number of setups at each locations will also be noted on this list, as will be the true sites to be used for filming.

This numbered locations list represents a very important budgetary element. As mentioned previously, a large number of locations can increase costs if they are not

LOCATION LIST AND SET UPS

1.	Living room	168
2.	Outside window	4
3.	Dock	58
4.	Front of the house	27
5.	Cockpit cabin cruiser	17
6.	Beach path	33
7.	Special	—
8.	Kitchen	6
9.	Beach near rocks	69
10.	Cabin cruiser (ext.)	6
11.	Bedroom	36
12.	Cabin cruiser (int.)	67
13.	Near beach path	15
14.	Near the house	3
15.	Near beach	5
16.	Point of land	14
17.	Beach (sand)	16
18.	Outdoor shower	8
19.	Terraza	17
20.	Back of the house	5
21.	Pick ups	44

located quite close together geographically. This list should be subdivided into two more groups: Location and Studio. On the Location side, space should be left for notes that will help the producer and director make decisions before or during scouting expeditions. If there are two or more possible sites, that also should be put down here; all this information will help greatly in cost calculation. For example, a nearby site may mean less transportation cost, but artistic considerations must be balanced off against the merely economic ones. A low-budget film does not necessarily mean the cheapest one, but does signify the best utilization of the available funds.

At this moment it is pertinent to insert a brief word regarding that often-mentioned abstract term *production values*. These values are intangible and reflect, to a great part, the good taste, or should we say, appropriate taste, of the producer or director. The script calls for a scene in a bar without specifying any particular type. There are several bars available, from a low-life dive to an exclusive club. The director realizes that one place reflects the true intent of the author of the screenplay while at the same time presenting a satisfying image on the screen. He chooses this one over the others, as it is, in his opinion, the best for the job because it fulfills his inner concept of the story.

Cost estimates of all these location rentals should be transferred to the budget section.

Starting with location number one, you may begin compiling the all-important breakdown sheets. Make up an original form and run off about 100 copies. Be sure to leave a margin at the left for perforations of the binding clip. The following information sections are needed on the breakdown sheet. At the top, list the location name with its corresponding number, leaving a blank for the number of pages and page identification.

The headings should contain the following information:

1. Page number from the script
2. Scene number
3. Interior or exterior
4. Day or night
5. Check off square
6. Eighth of page
7. Cast present (just initials)
8. A brief description of the action and/or dialogue

On examining the sample page, it will be noticed that all the information is condensed here, including all important 10/1-6, which means that this is location number 10, page 1 of a total of 6 pages. The total number of shots for this location is added up and transferred to the location list and the form sheet of the shooting schedule.

When finished, the entire script will have been logged according to the locations. Needless to say, the numbers should be checked off in the script to be absolutely sure that none is overlooked.

At the same time the location list is being made up, a cast list should be prepared if one was not included with the shooting script. At this point in the preparation it becomes necessary to decide in which order the locations will be taken, as well as the order in which the scenes will be shot. The great importance of this step is not apparent at first sight since it is not an artistic consideration at all — it is strictly financial. It deals mostly with contracts for players and, to some degree, their availability. To this end, the production office must make some quite difficult, but necessary, decisions. To help make these decisions, a diagram called a *cross plot* is made up (see Figure 2). To make up a cross plot, the cast list is consulted first, along with the breakdown sheets. If the leads appear throughout the picture, there is no problem, for they will be contracted for the entire production period. If, however, one or more of them appears sporadically throughout the picture, every effort should be made to bring their appearances together on consecutive days or weeks. This is especially important for contracting supporting players if their appearances are scattered. In these cases the salaries for the intervening days must be paid for, whether the supporting players work or not. If a lead player appears during only 3 weeks of a 4-week schedule, considerable savings can be made on his contract, entirely apart

	Living Room	Yacht	Beach	Dock	Path	Dunes	Cockpit	Rocks	Unused Days
Nick	X	X	X	O	X	O	X	X	2
Kitty	X	O	O	X	O	X	O	X	4
Banc.		X	O	X	O	X	O	X	3
Capt.		X	O	X	O	O	X		3

▶ ***Figure 2A***

	Cockpit	Dock	Yacht	Rocks	Dunes	Living Room	Beach	Path	Unused Days
Nick	X	O	X	X	O	X	X	X	2
Kitty		X	O	X	X	X			1
Banc.		X	X	X	X				0
Capt.	X	X	X						0

▶ ***Figure 2B*** *In this example the Xs represent working days at each location. The Os represent unused intervening days which must be paid. By rearranging the shooting schedule as in Example B, nine unused days are eliminated, effecting substantial savings in salaries of actors and expenses.*

from the fact that the player will have more free time for other engagements. Having separated the names of those who appear in all of the picture, we now do the same with those who work only 1 or 2 days. This leaves us with a group of players with different degrees of importance. Their appearances must be consolidated.

This is mostly a mechanical process. Take a large sheet of cross-hatched paper and make a list of the cast down on the left hand side. Then list all the locations on the top. With the breakdown sheets in hand, begin with location number one. Make an X in the square where the player and location coincide. When you finish with all the locations, you will have in front of you something that looks like a crossword puzzle, the correct interpretation of which depends on the number of appearances of a player, the number of locations where he or she appears, and the contractual obligations entailed.

There are no hard and fast rules on how the consolidation process should be done. A multiplicity of examples could be constructed, but none would be like the one that will appear on a new production since every motion picture is different (see Figure 3).

Shooting Schedule

#	Location	S	O	Days	Remarks	
3	Dock	58	C	2 1/2	Private Dock--South Padre Island, Texas	
5	Cockpit	17	B	1		
10	Cabin Cruiser	67	A	2 1/2	6 days	
6	Beach Path	33	B	1 1/2	Cameron County, Texas 170 --N. 2 Tracks	
9	Beach near rocks	69	A	4	207 --K.N.	
12	Near beach path	15	C	1		245 -- B.K. 250 -- K.N.
14	Near beach	5	E	1/2		310 -- K. 320 -- B.
15	Point of land	14	F	1		402 -- B.K. 473 -- B.K.
16	Beach (Sand)	16	D	1	9 days	6 Tracks
1	Living room	168	A	6	Interiors	
11	Bedroom	36	B	2	Beach House, South Padre	
8	Kitchen	6	D	1/2	Island, Texas	
18	Terraza	17	C	1	9 1/2 days	
4	Front of house	27	A	1	Exteriors	
19	Back of house	5	D	1/2	Beach House, South Padre	
2	Outside window	4	C	1/4	Island, Texas	
17	Outside shower	8	B	1/2		
13	Near house	3	E	1/4	2 1/2 days	
20	Pick Ups				Isla Mujeres	
21	Underwater				27 days total	

▶ ***Figure 3*** *In this example of a shooting schedule, S indicates the number of shots and O indicates the order of shooting.*

With the shooting schedule in hand, it's a simple matter to put down the number of working days after the name of each player on the cast list, which in this case should include everyone, with their contractual cost, all of which goes into the final budget. The rest of the information is transferred to the shooting schedule.

When the shooting schedule is made up it will look considerably different from the original location list. If you look at the sample shooting schedule, you will see that the locations are bunched together geographically. It will also be noted that the order in which they will be filmed is not the same as before. This is indicated by the letter *A* which is to be done first, *B* second, *C* third, etc., and is the result of the cross plot work that preceded this step. The number of setups is annotated and the number of days necessary to do the work determined and set down. Just how do we come about this last number?

Here is where experience comes into play. How does the director know that it will take two-and-one-half days to do 58 setups? An active low-budget director, on reading the script and looking at the number of setups, will say, "I have here 58 setups, some easier than others. At about 20 per day I can get it out in less than 3 days. Let's make it two-and-a-half." From comparison of many low-budget productions, the average result is about 22 setups per day. Of course, it's necessary to be flexible. If it is a new location that requires setting up lights, set decoration, or other time-consuming preparation, and it is to be used for only one day, the average will drop to around 17 or 18 setups. When several days' work and many setups are concentrated on one location, it's usually possible to save time and the average number of setups will increase to 25 or more.

It is entirely possible to film a picture in a shorter period of time by eliminating many close-ups and reaction shots, using only the master shot. This reduces the number of setups and, consequently, the number of shooting days, but the picture usually loses quality when this type of economy is practiced.

Now is as good time as any to talk of the *master shot*. It is not a long, a medium, or any other specific shot. It may start with a close-up, pull back to reveal other players, contain several movements, both of players and camera, recording various bits of dialogue — all taken in one shot. After a satisfactory take is made, the director moves in for the close-ups and reaction shots repeating the same action. This gives the editor more material to work with and usually makes for a much more interesting picture. Detractors of the master shot claim that it allows a producer or editor to change the content of a scene. This argument has no validity because the complete action is needed. Even though the master shot is not indicated in the script, you might see the possibility to shoot the sequence that way, saving more time than if you cut and replaced the camera for each close-up, then returned to the original shot. The director has the option to shoot a sequence in the way he or she sees fit, providing he or she does not change the basic intention of it and proceeds in a manner congruent with the economic structure of the film. Any important decisions or recommendations pertaining to the shooting schedule should be placed under "Remarks" in the right-hand column.

With this type of shooting schedule, there are a given number of setups at a certain location, such as 168 setups in the living room. The director realizes that he or she is going to be hard-pressed to obtain an interesting variety of angles, and camera and actor movements to provide visual interest entirely apart from simply telling the story. Accordingly the director will make notes, drawings, and diagrams in the script to take care of these shots.

When all this is ready, the breakdown sheets, location list, and shooting schedule should be fed into the copy machine, collated, perforated on the left-hand margin, and fastened together with a two-hole clip (see Figure 4). Six copies should be made up: one each for the producer, director, script supervisor, cameraman, and editor, plus one extra in case of necessity. The usefulness of this type of breakdown, apart from its original intent, comes from the fact that it is exactly the same size as the script, 8 1/2 x 11 inches, and that all concerned with the production can have a copy, whereas with other systems only one person knows what it going on. The knowledge of the shooting order is good for the rest of the people related to the production, allowing them to make a more efficient work.

You will have noticed that the shooting schedule is not a day-to-day enumeration of the shots to be made. If it were, it would be longer than the script and probably unmanageable. Leaving the development at this stage permits the director a flexibility in arranging the next day's work. This list of shots is given to the production assistant, who elaborates the call sheets and, in turn, delivers them to the assigned

Cabin Cruiser (Int)							
P	#	I/E	D/N	S	8	Cast	Action
10	44	I	D		2	K N	N goes to radio, turns it on, puts on headphones. K watches for a moment, turns to galley. MS.
10	45	I	D		2	K N	K takes bottle from locker (dialogue) N doesn't answer K. K comes to N, who waves her away. K turns in disgust, puts bottle on table, goes into cabin, strips upper. Puts on blouse Back to camera. MS. in depth.
10	46	I	D		1	(K) N	N takes of phones, turns radio off, goes to cockpit, looks out over water. Back to camera. MS.
11	51	I	N		2	K (N)	K is brushing hair, prettying herself (same profile as B#50) she finishes, turns directly to camera.
16	72	I	N		2	K N	N walks to radio, K passes to cabin. N standing puts phones to ear. He looks at watch, turns to see K undressing. N puts down phones, goes to K his back to camera. He takes her in his arms, turns her to him, kisses her. They fall on the bunk.
16	73	I	N		1	K N	CU Nick's hands carressing K.
16	74	I	N		2	K N	N and K on bunk, both nude, active kissing–lovemaking. MS.
16	75	I	N		2	K N	CU Kitty's face in ecstasy. MCU another angle.
17	84	I	N		2	K N	N is buttoning his shirt, tucking it into his pants. K is standing in transparent nightdress, rubbing her hair back in satisfaction. N moves to right bunk (his), rummages under or in drawer, removes small white box with plastic cover. N turns to face the camera with box in front of him. K is in the background

▶ ***Figure 4*** *The first column is the page number in this example of a breakdown. The second column is the scene number. The third indicates if the scene is interior or exterior; the fourth day or night. Check off is made in the fifth column when the scene has been shot; the sixth gives the scene length in eighths of a page. The cast column is wider, but initials only are used. A very condensed description of the action follows.*

WARDROBE LIST

KITTY

K-1 NIGHT CLUB
Tight-fitting sweater (no bra)
Tight pants
Matching platform shoes (bright color)

K-2 ARRIVAL SCENE (also under titles)
Hip-hugger pants
Deep V cut upper
Tennis shoes
Light colored sport glasses

K-3 FIRST VISIT TO BANCROFT'S HOUSE (also second visit)
Blouse (short, revealing)(white)
Sport shoes
Tight pants (yellow)

K-4 NIGHT SCENE (cabin cruiser Argus # 84)
Night dress (transparent)

K-5 FIRST SWIM SCENE #111
Bikini (yellow knotted string)

K-6 UNDERWATER - BEACH - SHOWER SCENES
Bikini (red knotted string)

K-7 SHOWER - DOCK ARGUS
Large white bathrobe

K - 8 DANCE SCENES # 253-B
Tight pants (hip-hugger) (light blue)
Revealing blouse (blue)
Bolero-type jacket
Large silk scarf

K - 9 BURIAL SCENE
Silk handkerchief (medium size, neither light nor dark)
Dress (light tan, button-down front)
Same shoes as in K-3

K - 10 BEACH SCENE # 468
Blue jean shirt top
Blue jeans

K - 11 FINAL SCENES #489 on
Blue jeans
Tight-fitting sweater
Matching platform shoes (bright color)

actors and technicians. The script supervisor appreciates having more material to work with. The cameraman armed with this breakdown and shooting schedule knows exactly what is ahead and can be prepared for it.

A good example of the use of this breakdown and shooting schedule is the picture *Deadly Reef*. It had a 27-day shooting schedule and the director was able to come up one full day under schedule, which made the producer and all concerned quite happy. Four days were set aside for underwater photography. This consisted of 16 shots from prepared storyboards and was contracted out to a specialist who shot it in two days, using doubles.

It spite of the fact that everything was programmed, what finally became one small five-second pick-up shot of a small, light plane against a bare blue sky was a white piece of leader film in the work print for two months, until the director personally went out and shot the footage. A stock shot, you may say? Library material? None were satisfactory. When you start a picture, do not count stock shots unless you have already viewed them and have the negative and work print in hand.

Now we will explain the paperwork — all important paperwork. *The wardrobe lists* are part of this work. It is necessary to go through every page of the script, scene by scene, to determine the clothes that each character is wearing (see Figure 5). This is absolutely essential and copies must be given to the producer, director, assistant director, script supervisor, makeup person, and wardrobe person. They are vital in the avoidance of *mismatches*, besides being necessary for the budget and purchase of the items. Some actors prefer to supply their own wardrobes for contemporary films.

Wardrobe--Bancroft			
Costume	Set	Time	Scenes
Number 1	Int. living room	Day	1–5
	Ext. house	"	6
	Ext. beach path	"	11–25
	Int. living room	"	30–39A
	Ext. front house	"	41
	Int. living room	"	42
	Ext. beach	"	42A–42B
	Int. living room	L. Afternoon	47–50
Number 2	Int. living room	L. Afternoon	55–68
(changes	Int. living room	" "	79–82
shirt)	Int. bedroom	" "	83–95
Number 3 *Ski pajamas*	Ext. near beach (stalks Nick)	Night	98–99
Number 4 *diving* *suit*	Ext. swimming near dock surprises Kitty	Day	110–120

▶ ***Figure 5*** *Sample page of wardrobe use. This particular wardrobe is for Bancroft.*

PROPS LIST

INT. LIVING ROOM - BANCROFT HOUSE

PROPS:
Typewriter
Reference books
Other books
Reams of paper
Carbon paper
Pencils
Pens
Erasers
Wastepaper basket (filling available)
(201)
Folders
Magazines
Binoculars
Ashtrays
Beer & glasses (tray?)
(from kitchen)
White wine and bottle
Radio??
Petate (or rolled mat)
Plates and utensils from kitchen
Bottle of cointreau (in sideboard0
Small cointreau glasses (3)
Small shot glasses (3) (253C)
Square bottle (264)
Coffee pot
Coffee cups and saucers
(from kitchen)
Box of palillo cigars
Bancroft's glasses
Bancroft's pipe, tobacco, & matches
Flashlight or gas lamp
Balled-up papers)
Half-empty glass) - (#224)
Wastebasket filled
Flashlite (224)
Kitty's record player)
Kitty's records) - (254A & #477)
Kitty's scarf (254A for 256)
Bancroft's lunch (small) (300)
Harpoon and underwater gear
Bottles: antiseptic, alcohol
Bandages, adhesive tape
Towels
Kitty's chain and crucifix (#380)
Bed sheet (382)
Shovel
Small wooden cross
Small book (388)
Breakfast for Kitty and Bancroft
(#458)
Kitty's nail polish, buffer etc.
Kitty's suitcase (closed and filled)
(533)
Kitty's (left behind) bikini (#543)

DRESSING:
Oil painting of The Reef
Poster of bullfight
Writing desk
Louvered shutters
Light curtains
Photos on Walls:
with tiger or lion
with elephant
Mounted fish
Stuffed animal heads
Animal skins
Guns in cabinet
Shark jaws
Fishing nets
Fishing tackle
Spears for fishing
Underwater tanks
Underwater masks
Barometer
Peg to hang binoculars
Sofa
Pendulum clock
Sideboard
Table and chairs
Overhead lamp
Plants
Other lamps??

NOTE:
Nick's scalp cut (#369)
Blood) - (#364)
Sand)
Sweat bottle
Rain effect (421)
Lightning effect (#423A)

WET DOWN ACTORS (#435)
Need towels

This is an advantage, but the apparel should be approved by the director, or he or she may end up shooting some very strange costumes indeed. Usually it's better to provide the wardrobe.

In any case, on low-budget films the clothing most be purchased off the rack of moderately priced establishments. If there is any danger to the wardrobe or should it require a lot of use, be sure and double or triple those items. It's a small price to pay for peace of mind. Some players are conscientious about the care and correct use of their wardrobe — others are not. It's the combined job of the script supervisor, wardrobe person, and the assistant director to see that the actors are wearing the correct apparel at the start of every scene.

Now we come to the *property list*. On a recent low-budget feature, a pendulum wall clock, binoculars, a portable typewriter, and a special hand-worked water pump were all called for in the script, as well as innumerable other props. The list was made up and all the items acquired before the production shooting. One time, a big production with 40 technicians and over 15 players stalled for several hours because a certain type of pistol had not been provided at the proper time. Minutes run into hundreds and thousands of dollars when everyone is idle. Try to avoid these delays by having everything on hand.

The wardrobe and property lists should be priced out and transferred to the budget section. If some items are to be made on special order, such as dance costumes or wedding dresses, the sizes of the players should be available and arrangements should be made for fitting appointments if required. It is interesting to note that costume and prop rentals are available, but prices must be compared before initiating acquisitions.

This chapter also presents probably the most important points of production office work, aside from the budget, which will be considered in the next chapter. You will have observed, perhaps to your disenchantment, that there is very little improvisation in the production of a commercial film, yet there is a constant presentation of choices to be made concerning locations, wardrobe, properties, set construction or rental, color schemes, styles, and so forth, all before the actual shooting begins. This is the difference between conscious preplanning and on-the-spot improvisation. One provides definitive artistic choices and expedites the making of the film. The other may lend an air of authenticity to a picture; however, there are doubts about this. Most of the great artists of history prepared drawings or so-called cartoons, upon which they based their definitive large-scale productions, and they certainly were not fettered by this process. It is worth some meditation on the part of those who would decry thought and preparation in favor of off-the-cuff shooting.

SUMMARY

- Make a list of locations (location and studio).
- Make up a complete list of the cast.
- Make up breakdown sheets.
- Make up cross plots for scheduling.
- Construct shooting schedule with estimated production time.

- Make up individual wardrobe lists.
- Make up prop lists.
- Copy and bind breakdown sheets and shooting schedule.
- Transfer estimated costs to the budget section.

3

Financing the Low-Budget Feature

Probably one of the most challenging and difficult of all projects is the financing of a motion picture of any kind. The major film producers have entire departments worrying about revolving credits, major bank loans, advances, payments from the big exhibitor chains, overseas remittances, devaluation in many countries, overblown salaries and percentages to overrated stars, and union demands. In addition, they must faithfully fulfill their obligations to their stockholders, just like any other business, accompanied by the ever-present fear of a takeover. To keep the monster rolling, they must urgently finance and produce a certain number of viable commercial pictures every year. Of course, they have a giant machine utilizing multimedia publicity to do just that, produce and market pictures.

Where does the low-budget producer fit into this picture? Let us take an overall view of the business aspect of the industry. There are three categories of film productions: the majors, the minors, and independents.

The great studios of Hollywood, of which there are eight, are called the *majors.* The budgets for their productions run from \$20 to \$60 million. Smaller production companies, usually contractually related to the majors, are known as *minors* and operate on budgets that vary from about \$8 to \$20 million. The independents, who are numerous, have budgets that are considerably restricted, starting from sums as low as \$200,000 to \$5 or \$6 million.

There is considerable overlap between the last two categories, apart from the so-called independents, which they are not, who work directly for and with the majors.

The marketplace is divided roughly into 90% for the majors and 10% for the independents. In the same way, the financing reflects these percentages.

Some independents, who may also be classified as minors, are stronger and represent about 2% of the mentioned 10%. Another 2% are poverty stricken, with the remaining 6% representing the majority of low-budget producers. This will give you an idea of where you stand without being burdened with the precise sums to the last decimal point as stated by the Department of Commerce. The New York newspaper

Variety publishes, from time to time, very accurate figures as to the economics of the industry, as well as up-to-date information on worldwide production, distribution, and exhibition. It comes out every Wednesday — get it and read it. The reporting is based not on gossip but on statistics.

Where did the independents get their financing? The use of the past tense is intentional. If a company owns a picture, or pictures, made in the past, it has assets. If it owns only a script, that is an asset.

Production begins with some sort of base. The construction and utilization of this base is our next subject and there are many ways to do this, some right and some wrong.

Personal or Family Finance

Personal financing of a film project is a method of film production that requires enough cash on hand to finish the principal photography. It is important to remember that an unfinished picture is a total loss. Of course it would be desirable to have sufficient money to complete the entire production; however, if the photography is finished, it is always possible to hold up postproduction for a reasonable period of time until further funding is available.

Mortgaging other assets, such as real estate or property, is a common way of getting cash. Taking loans against a prosperous business other than films is also a useful way to acquire funds. Personal loans are sometimes utilized, but some real collateral is always necessary. A picture yet to be filmed or in the process of being filmed cannot be considered as collateral.

When you read about bankable stars in trade publications, the writer is not talking about low-budget pictures. The salary of such a star would finance several pictures of the type described in this book. Such salaries are out of reach.

Bank finance requires a number of guarantees usually unobtainable for an independent producer: signed contracts of the stars, distribution contracts, a completion guarantee, special laboratory contracts, a completely detailed budget, a financial statement of all the risk takers, and a "name" director.

The one or two banks involved in motion picture finance are primarily interested more in being repaid than in the merits of the project, and base their judgment on these considerations. Large independents, working under the credit umbrella of major studios, can possibly fulfill all these requisites. This route is not feasible for low-budget producers.

There is a subject that must be discussed here because many persons new to film production count on unsecured personal loans from parents or relatives. Do not take personal family loans. There are many reasons against doing it and none in favor. The motion picture business is a risk business — from the smallest production to the largest. Unless your relative has lots of money and doesn't mind putting it into your project, you can easily expend your grandparents' life savings and have nothing to show for it. On the other side, they may make a fortune, but it's best not to gamble.

Cooperative Ventures

For aspirants to filmmaking who have many friends all working toward the same end, the cooperative can be a wonderful solution to a complex problem. Larry will direct, Joe will be the cameraman, John will edit, Jim will do the sound, Mary the script, and Jane the wardrobe.

Around the coffee table it sounds fine: a rented camera, some lights, the star of the little theater group as the lead, his or her friend as the second lead. All these, if properly developed, will produce a picture.

Cooperative ventures, both in consumers and production areas, are praiseworthy. The cooperative system is morally and economically sound. The structure is quite smaller to the corporation and limited partnership, with one very important exception: the members, who are equivalent to the shareholders or limited partners, participate and work directly in the production and perceive financial rewards equivalent to that participation. A similar financial arrangement was the basis for a prominent film company founded in Hollywood in the 1920s.

The legal structure, which is governed by state law, must be set up by a competent lawyer familiar with cooperatives. The institution is presided over by a duly elected president aided by a secretary, treasurer, and other officers, all serving for a definite period of time, usually a year. This is an important aspect of cooperative organization. Once elected, the president is free to govern the institution as he or she sees fit, within the rules and regulations of the institution.

A cooperative may buy, rent, or lease equipment, carry out credit and finance operations such as loans (with collateral, of course), but may not issue stock. Membership may be increased if convenient. Many cooperatives in extremely diverse fields have been enormously successful. There is even an international organization representing cooperatives from all over the world.

Having decided on the project and who is to head it, the members should make every effort to collaborate toward its successful termination and set aside all their personal differences, subordinating their individual egos to the common good. There exists historical counterparts to cooperatives in Hollywood, some that were quite successful and some not so, although they were usually instituted as corporations.

It is not necessary that everyone needed in a picture be members of the cooperative; actors, technicians and personnel may be contracted as needed, as such an organization is flexible enough to permit practically all corporate functions.

If you and your friends are looking for a way to enter the motion picture field as a group, investigate the world of cooperatives. I'm sure you won't regret it.

The key to a successful cooperative consists in the election of a president who is intelligent, intellectually honest, and highly motivated.

Simple Partnership

Partnerships are not practical for film productions because a partner is responsible for the debts incurred by the other partner. Legally it is possible to proceed against the personal assets of either or both to cover obligations related to the production. This form of partnership is, for all practical purposes, obsolete. It is convenient to note here that a person or company can be sued for not fulfilling contractual obligations, even though no real debt exists.

Limited Partnership

The limited partnership is a corporate-type structure set up between the general partner (producer) and a group of limited partners (investors), delineating and limiting financial responsibilities to the extent of their participation. It permits the producer to carry out his or her function, and at the same time stipulates the manner in which profits will be divided among the investors after expenses are paid out.

This type of financial structure allows the entrepreneur to organize and obtain funds to use for low-budget productions. The limited partnership is, however, restricted in comparison to a corporation where stock may be issued and sold, first by private offering and second by a public offering after registration. The limited partnership is economical to set up and can be amply recommended.

Incorporation

The best, most complete, and most versatile of all business organizations is the corporation. The corporate structure also provides the easiest way to raise capital for your film venture by making a private or public offering of securities, and allows for a continued existence and participation in production without limitations on expansion. Although many corporations are formed to produce a single picture, most of them continue indefinitely. All types of financial manipulations are possible: common stock, preferred stock, warrants, bonds, debentures, etc., as well as intrastate offerings — private and public without registration if the operation is carried out under Regulation A of the Securities and Exchange Commission.

Corporate law is quite complex, but there are many books available on corporate structure and finance, even on how to do it yourself. If you are thinking of forming a corporation, take the time to read up on the subject at the public library, and if you have access to a law library, you can submerge yourself completely or at least get as much general knowledge as possible regarding corporations.

Some information that you vitally need is not in the books. One concern is the cost of incorporation. This can run from about $300 for a do-it-yourself job to $25,000 for the services of highly paid corporate lawyers. By shopping around, you can incorporate for somewhere between $1000 and $1500. Be sure to get a lawyer familiar with corporate law. Just any lawyer won't do; besides, it's always good to have counsel to refer to in case of need, but make your fee arrangements in advance so you will know exactly where you stand. By all means, be aware of your responsibilities, otherwise you may run afoul of the Federal Regulatory Agency, the Securities and Exchange Commission. Don't let this frighten you off. Incorporation is the best way to go for financing a low-budget production.

Another important fact concerning corporations is that you are in control only when you own or by proxy have a majority of the voting shares. One well-known successful mini-major producer, a lawyer himself, recently consummated a merger with another company. In theory, everybody was going to benefit from the deal, which looked very good, but in a few short months the old-line producer of low-budget pictures found himself booted out of his office onto the street! By some peculiar oversight he had lost control of the newly merged company. Of course he had become a millionaire, holding his own stock, which he sold off as soon as he could, and now is starting over again.

There are a long list of advantages to be gained by incorporating the production company. They are

1. A corporation can sell shares and raise capital.
2. Fiscal responsibility is limited.
3. The corporation can obtain credit.
4. The corporation can solicit loans.
5. The corporation can acquire assets.
6. The corporation can expand to a national level.
7. There is personal immunity from lawsuits.
8. Sale of securities can be made by stockbrokers.

The selection of officers for the corporation is of great consequence and all the members must be competent and dynamic. It is recommended that the treasurer of the corporation be an accountant with knowledge of the stock market. These matters, such as securities, offerings, and writing a prospectus, seem far from the motion picture field but are the real base of the industry as, today, more films than ever are being financed through sales of stock in one form or another.

One final admonition: When offering your project, do not make claims or assertions that can be misconstrued as such in regard to the results of the project.

When a production company has a concise, dynamic plan for progress it will succeed in making and marketing pictures. The corporate structure provides a solid base from which to work. Investors are always available for good profitable projects.

Going back to the field of personal finance, we should take a brief glance at the subject of grants. There are many institutions doling out small grants consisting of $1000 to $25,000 for filmmakers, some of whom have been helped along with these funds. Not a month passes without some article being published about grants that increases the hopes of novice filmmakers. The sad truth is that a grant would be a proverbial drop in the bucket of a commercial producer. If the sum were enough to cover the negative cost, it would scarcely be worthwhile. But, in the case of a 16mm documentary of less than 30 minutes, it could be of help. Time spent in soliciting grants could be better employed in search of other sources of capital.

Many countries of the world have subsidies for film production, such as in England, Germany, France, and Italy, but most of these monies are tied into specified characteristics of production, which rarely, if ever, apply to a foreign company initiating production. This is not to say that these subsidies don't help some companies — they do — but they don't help the ones that need it most. Besides, there is always a long list of local talent in queue ahead of you. It is nice to know of these subsidies. Now let's return to reality.

Another pertinent aspect of personal financing is fiscal responsibility. If you have the money and go about making a picture without a corporate structure, you are personally responsible for all debts, liens, and lawsuits incurred at any stage, before, during, or after production. You may be slapped with a completely unfounded lawsuit alleging plagiarism because someone registered a similar screenplay 20 years ago. That is why corporate structures have their advantages.

Condensed Budget & Program of Dispersements			
Item	Sum	30 Days	90 Days
Producer	$15,000	$7,500	$7,500
Story	7,500	7,500	
Director	15,000	7,500	7,500
Cast	25,000	25,000	
----------	----------	----------	----------
Technicians	$15,500	$15,500	
Construction	6,500	6,500	
Location	40,000	40,000	
Wardrobe	4,000	4,000	
Make Up	1,500	1,500	
Film Material	35,000	33,000	$ 2,000
Music	15,000		15,000
Laboratory	40,000	20,000	20,000
Sound	12,000	1,000	11,500
Editing	9,500	2,500	7,000
Titles	3,500		3,500
Rental Equipment	12,500	11,500	1,000
Insurance	3,500	3,500	
Pre-Production	3,500	3,500	
Preparation	1,750	1,750	
Miscellaneous	5,000	2,500	2,500
Totals	$271,750	$194,250	$77,500

1-4 (above the line) $62,500	30 Days $194,250
5-20 (below the line) $209,250	90 Days 77,500
Contingencies (20%) $54,350	Cont. 54,350
	Total 326,100

▶ ***Figure 6*** *The first column of this preliminary budget represents a rough estimate of the total cost of the picture. The second indicates sums expended during the first 30 days. The third represents postproduction costs, up to and including the answer print.*

Investors are, by law, advised in the prospectus of a corporate offering that there are no guarantees, and they purchase the shares with that knowledge.

Note: Any opinions voiced in this chapter must not be construed as legal advice; that can be obtained only from a lawyer.

Preliminary Budgets

The first time you look through an itemized budget for a full-length feature film you will be taken back (see Figure 6).

This is not what the teachers taught you at school. This is not art, this is drudgery! It is, but do you think the great director or producer escaped from this work? They certainly did not. I can, from personal experience, assure you that you, too, will pass sleepless nights with this part of filmmaking. It must be done and done thoroughly because this will determine whether you can make the film or not — and whether you can finish it or not. Take the budget as a voyage of discovery as it must be done; enjoy its intricacies. When you look over the items you will be astounded at the things you haven't taken into account and that must be considered. Did you ever stop to think who paid, and how much, for the block of marble Michaelangelo used to sculpt his monumental David? Who paid for the paints and canvas that Velazquez used? Today we stand and admire the results but the production costs and problems never enter into our minds. We are not told the whole truth about production, not so much because of secrecy, but simply because it's not pertinent to our enjoyment of what we are viewing, however important it was to the person who made it.

A motion picture before, during, and after its production usually goes through several budgetary processes, beginning with the significant preliminary and ending when the accountant lays down the definitive final budget in front of the producer who, at that point, is usually both worried and delighted — it's finally finished — or is it?

The preliminary budget usually sets down the overall guidelines for a production. Before this cost-estimate can be obtained, it is necessary to make some basic decisions that vitally affect the total. To set down the results of these decisions, a catalog of accounts must be constructed. The usual form consists of a page containing the condensed figures under approximately 25 headings, followed by about 15 pages of itemized accounts subdividing these headings. Some production offices use printed forms available from specialty printers, while others make up a catalog tailored to their specific needs. The results are almost identical in any case.

Above-the-Line Costs

On consulting the boxed text, you will notice a line drawn across the page, dividing the first four items from the rest. These above-the-line costs are subject to wild and varied fluctuations, which is the real reason for their separation from the others, the prices of which are conventional and relatively stable.

The first four items are story, producer's, director's, and actor's fees. We will consider these and other items that basically relate to the production of low-budget

CATALOG OF ACCOUNTS

1. Story purchase, treatment, and script
2. Producer
3. Director
4. Cast

---

5. Extras and doubles
6. Production personnel
7. Operating personnel
8. Operating expenses
9. Postproductions expenses
10. Laboratory and film
11. Main and end titles
12. Rerecording expenses
13. Music, arrangements, and copy.
14. Location and transportation expenses
15. Studio rentals
16. Retakes and tests
17. Publicity, promotion, and printing
18. Insurance policies and premiums
19. Fees, taxes, and licenses
20. Employer contributions
21. General office expenses
22. Preproduction expenses
23. Miscellaneous expenses
24. Contingencies

All receipts should be categorized, i.e., makeup purchases should carry 8F1 written in red ink with a circle around it. The accountant will then know where it belongs.

features. Most of the aspects concerning the story were taken up in Chapter 1. However, it's necessary to mention a few relative points.

Rarely does a small independent producer purchase the motion picture rights to a novel. The rights for a major novel could cost more than the budget for the entire picture. Oddly enough, screenplays can be had by the hundreds; the only difficulty is getting your hands on one that can be produced for very little money and that has sufficient impact to give your feature a boost at the marketplace. Nowadays that means enough sex for an R-rating, without falling into the X-rating pit. Almost any degree of violence is permitted. It's a curious comment on our times that destructive impulses are condoned whereas the essentially creative impulse of sex is castigated.

If you don't have a low-budget screenplay, ask any Hollywood agent and you'll have reading matter for the next 10 years.

Set down the cost of the screenplay and related items under the appropriate heading.

The second item above-the-line usually refers to producer's fees. We must remember that the word *producer* fits two categories with considerable subsequent confusion. We would all like to have a real live producer — i.e., the one who puts up the money for the entire production. These are naturally a little scarce. Producers of this sort rarely receive a fee unless they actively work at production.

The executive or assistant producer is a holdover from the days of the giant studios when there were A pictures and B pictures, assigned according to category. Today associates, investors, boyfriends, and girlfriends are called producers, without rhyme or reason. On low-budget pictures the person who does the actual production work or supervises it fits into this category, regardless of screen credit.

Some advice on producers is obligatory: Stay away from executive producers or production assistants from the big picture field. You may hope to benefit from their expertise, but usually it doesn't work that way. They are accustomed to ordering a thing done, rather than doing it themselves. They really represent an additional burden on a low-budget picture. Look for an active, energetic person with experience who can get things done with alacrity and who is capable of turning out all the production office work.

The director of low-budget feature films is a completely different person than the director of a multimillion-dollar epic. It's like comparing a field commander out on the front with his troops in battle to a staff officer who works with a small group of men while his subordinates carry out the rest of the duties. Our director is an intensely practical person who knows that there are 20 to 25 setups every working day, six days a week, with an average of three takes per setup. The director must know what he or she wants right now. Good low-budget directors are just as scarce as good high-budget ones. The going salary is from $10,000 to $30,000 per picture. If the director supervises the editing, a higher salary is required.

The salaries of actors represent the fourth and last group of above-the-line costs that can quite literally make or break you. More than any other item of expense, the salaries of the players represent the difference between low- and medium-budget films.

The Screen Actors Guild has several books similar in size to the telephone directory, crammed with the names of actors of all ages, colors, and sizes. The books do not specify salaries. Many of these players specialize in low-budget film work. The best way to find them, aside from personal experience, is to use the services of a medium-sized agency. Why medium-sized? The large agencies are not really interested in serving low-budget productions; there just is not enough money produced by commissions. The small agencies may be eager to serve the producer, but may not have sufficient actors signed up to provide the adequate choice needed to fill the casting need. It's not necessary to contract all the cast from any one agency. However, it may be possible to package the whole cast from one source. Tell your agent, or agents, what you need and what you expect to spend, and they will present you with a selection for interviews or readings. A good agent can usually cast a low-budget picture in less than a week. Stills, video cassettes, and film clips can all help in making the correct choices.

Do not fit your budget to the actors' demands, but fit your choice of actors to the available money. This is the rule in our category of pictures. The chapter on "Actors and Agents" will provide a more complete exposition of salaries and terms for players.

When added together, the above-the-line costs on low-budget feature films should be somewhere near 25 percent of the total cost of a picture.

Below-the-Line Costs

All of the remaining items, numbered from 5 to 24, are considered below-the-line costs that are not susceptible to negotiation except for economies in purchase, discounts, or package deals. A catalog of accounts is of little value unless the corresponding numbers are available under each heading. Compiling this cost information is the principal job of the production office. The production manager or assistant producer phones, writes, calls on suppliers, and generally elicits all the information necessary to make up the initial study.

Under item number 10, for example, which deals with film stock and contains several subheadings, the office will annotate the latest prices of negative film stock, sound negative material, and intermediate positive material. Low-budget directors usually try for a three-to-one ratio of film stock, which can run as high as seven- or eight-to-one on larger budget pictures. Since this material comes in 400-foot rolls, about 61 are needed to make up 24,400 feet of negative for a 90-minute picture, which is 8100 feet in length.

Another example that comes to mind is item 14, Location Expenses. It contains a subheading, 14D—Meals. After you have calculated from the breakdown and shooting schedule the time required to shoot the picture, ascertain the number of personnel, including players and technicians. Multiply this number by the number of meals to be provided by the number of days of work in the picture. Then get the best price possible from restaurants or caterers, and calculate the total costs. You will notice on the catalog of accounts a number such as 8D5—Grip truck rental. Every

ACCOUNT CATALOG

Account 8D	Rental grip equipment	8F5	————
8D1	Rental equipment	8F6	————
8D2	Purchase equipment	Account 8G	Wardrobe
8D3	Diffusion material	8G1	Wardrobe rental
8D4	Filters, acetate, glass	8G2	Wardrobe purchase
8D5	Grip truck rental	8G3	Dry cleaning, ironing
8D6	Mileage, operating expenses	8G4	Rental sewing machine, iron, etc
8D7	————		
8D8	————	8G5	Cloth, thread, materials.
Account 8E	Prop rental	8G6	————
8E1	Prop rental	Account 8H	Miscellaneous
8E2	Prop purchase	8H1	Coffee, tea, water sodas, snacks
8E3	Purchase paints, furniture, etc		
8E4	Rental furnitur	8H2	First aid material, medicines, etc.
8E5	Purchase plants, flowers, etc		
8E6	Rental miscellaneous	8H3	Telephone service on location & set
8E7	————		
8E8	————	8H4	Rental special equipment effects
Account 8F	Makeup and hair		
8F1	Makeup purchase	8H5	Purchase special equipment effects
8F2	Hair and wig rental		
8F3	Beauty shop expense	8H6	————
8F4	Barber shop expense		

rental of trucks, cars, etc., on this account should be accompanied by a receipt and written on this circled 8D5. This makes it easy for the unit accountant to categorize all the current expenses without running to someone to ask. Everything must be kept in order by using the catalog and numbering all receipts. Accountants new to the motion picture field will have to familiarize themselves with the terminology.

One of these items pertains to the Screen Ratio and the Negative size. At the present time there are only three negative sizes available for low-budget theatrical feature films: 35mm, 16mm, and Super 16.

The ubiquitous presence and relative cheapness of 16mm equipment brings to mind thoughts of great savings on film stock and equipment rental.The fact that the camera and editing equipment are available appears, on the surface, very convincing, but let's lift off the cover and look a little closer.

Today screen formats are fairly well defined. Some years ago, 1.33-to-1 was the screen ratio for all theaters, and the standard aperture plates of all cameras are still based on this concept, the reason now being the ratio of the television screen, which it approximates quite well. When we look at theater screens, however, we see that the accepted ratios worldwide are from 1.66-to-1 to 1.85-to-1. What does this mean to us in relation to 16mm? It simply means that, as the screen ratio goes up, we use less and less of the original negative area to fill that enormous screen out there in that drive-in! Of course some area loss is experienced in 35mm, but due to the larger original negative size, it's of much less importance.

You will say that you have seen or at least heard of a picture where standard 16mm film was blown up to 70mm for national release. Yes, it was. An enormous amount of film was shot at a rock concert using multiple cameras, then edited down to feature length, accompanied by special sound tracks. The resulting grain and fuzziness was compensated for by the content, both of sound and image, and the picture was accepted. This system will not work on a regular theatrical feature.

Leaving to one side the technical reasons against blow-ups, let's look at the economics of the problem. Number one is a sword with two sharp edges, namely the fact that 16mm negative, or positive for that matter, must be edited into A rolls and B rolls, due to the fact that there is no space between frames for invisible splices. Now A roll/B roll printing does make for the easy introduction of opticals but, if you aren't going to use them (as they are mostly outmoded), there is no saving involved.

If you have used Eastman color 16mm negative, you must go to a 35mm interpositive on blow-up or to an internegative process, both of which are expensive. If you shoot Ektachrome ECO positive film, you have to put up with low film speed on interior sets. The blow-up is cheaper than the other, but is still about $1.50 per 35mm foot at the time of this writing, and prices are still going up. In the end, blow-ups come out the same as having shot in 35mm, except for the fact that you have, with some sacrifice in quality, deferred part of the final cost for some months, the only other difference being in equipment rentals, which are a little less than in 35mm.

There is another reason that is neither technical nor economical: Professional people psychologically prefer 35mm to 16mm; whether they express it or not, your actors think less of a picture shot in 16mm. Distributors and exhibitors are not happy if they know that the picture was shot in 16mm.

Another plus for 35mm is the ease in editing, because both the frame and the film are larger, making it more manageable. The big gauge wins, hands down.

Now for Super 16. In an effort to use cheaper negative material and equipment, the idea came up to take advantage of the unused sound track area of the 16mm frame, thereby reducing the losses in ratio on blow-up and adding more original negative area. It sounds good, but the camera has to be modified. The optical center of the lenses must be moved, requiring a change in the camera front. Some 16mm lenses won't cover the new area, so new lenses must be obtained. For the sake of safety on any picture, two cameras must be on hand. With Super 16 this means two modified cameras. Another important matter concerns the lab processing your negative. It must abe set up for developing and printing Super 16. This means that all of the hundreds of rollers in the automatic processing machines must be relieved or turned down on the sound track side so the negative or positive will not be marked or scratched. Super 16 is haunted by these defects. Consult your lab before proceeding far along this line. They can advise you if they are prepared to carry out the work and detail the problems involved.

So 35mm it is, the oldest, the best all-around negative size with rental equipment available everywhere, easy to work with, and at the very end, cheaper. Don't be misled by superficial charms, but think the matter all the way through. Any sacrifice at the beginning will pay off in the end.

Petty Cash

One of the evils we must live with is petty cash. A sample sheet on the following page illustrates how this varied and useful item must be cared for, as it has to be held in control. The sum usually runs from $500 to $1000 a week. At the risk of seeming parsimonious, we recommend that the accountant check it out and question the items just like any other account. Little vices sneak in that can cost the production quite a bit of money before the picture is finished. This sounds stingy, but this is Gower Street, not Century Plaza West.

How Much Money Is Really Needed?

When the preliminary budget is completed as minutely as possible, we arrive at the final item — contingencies. Traditionally this item was calculated at 10% of the total budget and was used to cover the inevitable small overruns in costs that occur on every picture. "On-time and on-budget" used to be the passwords for success. We are indeed fortunate to be able to quote the first part of the saying in today's world. The reason is simple: inflation, which, in spite off all efforts, remains high worldwide, seemingly built into the existing economic structure. We must take into account that inflation in the motion picture world is not expressed in relation to the general economy, so the rate is much higher. Equipment prices have soared hundreds of

PETTY CASH

EXPENDITURES

	Received ___date___		$200.00
	Received ___date___		$100.00
	Total on hand		$300.00
1.	Paper, pencils, records	$ 7.94	
2.	Liquor, ice, beer	8.46	
3.	Alcohol, merthiolate, adhesive tape, bandages, etc.	3.76	
4.	Cigars	.87	
5.	Varnish, paints, brushes, wood	11.94	
6.	4-inch spikes	.38	
7.	Multitester	9.40	
8.	Tobacco	.49	
9.	Gaffer tape	6.50	
10.	Advance to helper	5.00	
11.	Soft drinks and ice	8.40	
12.	Film for stills	10.19	
13.	Sugar	.71	
14.	Soft drinks, ice	5.04	
15.	LD call to Mexico - J. Randall	1.63	
16.	Parking Brownsville (no receipt)	.25	
17.	Copies of Governor's telegram, chalk	1.83	
18.	Lamps, lens tissue, still material	38.63	
19.	Black tape	.86	
20.	Socket clips (clamps)	24.18	
21.	Lamps, clamps	22.60	
22.	Lunch	3.68	
23.	2 days' work Asst. Production	40.00	
24.	2 telephone calls Matamoros	1.63	
25.	Haircut	10.00	
	Total spent	$224.37	
	On hand ___date___	75.63	

percentage points in the last few years. This is one of the reasons so few hard numbers are expressed in this book.

So what should we do about contingencies? The best way is to double the traditional figure — make it 20% and you will be right on: 10% for inflation and 10% for emergencies.

With this in hand your budgetary work is almost finished. All that remains to be done is to calculate an outlay or cash flow sheet, apportioning the sums needed by the end of principal photography, the delivery of the answer print, and deferred

USEFUL FORMS

1. Budget Sheets — Condensed
2. Budget Sheets — Detailed
3. Forms for Independent Contractors
4. Employment of Day Players
5. Release Forms — Physical Likeness
6. Release Forms — Property
7. Freelance Contract — Actors
8. Camera Report Forms
9. Laboratory Report Forms
10. Forms for Sound Mixers
11. Forms for Social Security, Withholding, etc.
12. Receipt Book
13. Time Book

READY-MADE FORMS.

There are many forms available for the independent producer. These should be taken advantage of whenever possible as time-savers. If a needed form doesn't exist and a number are required, make up an original and have it photocopied. When used in large quantities, printed forms are preferable and are much cheaper.

payments. In the first column, list the same items as the catalog of accounts, in the second the total calculated expenditures, and in the third the sums required at the end of the 30-day shooting period.

You will notice that the sum calculated for contingencies is separated since there is no known way of apportioning these happenings.

The bottom line of the second column is the key to the whole picture and for your sake, for the picture's sake, be sure you have the cash on hand to cover this figure before you begin. There is nothing as demoralizing as running out of money at some stage of filming and having to shut down the production. There is an almost irresistible temptation to start filming after so many months of preparation, but remember, if you finish the principal photography you have a picture. You can always wait on postproduction, but if you can't finish shooting, you haven't a picture, period!

SUMMARY

- There are three categories of film production houses: the majors, the minors, and the independents.
- Ninety percent of the market goes to the majors; 2% goes to the minors; the remaining 8% goes to the independents.

- Using personal money to finance the film is risky at best—and impossible most of the time. Do not use a relative's life savings to finance a film.
- Cooperative ventures can be a viable method for independent films. However, it is important to appoint one person in charge to make most of the decisions.
- Partnerships are hazardous. Partners are liable for each other's debts and legal obligations.
- Limited partnerships are, however, a different story. The partners are limited in their obligation and participation. It is more limited than a corporation, which can sell stock.
- Corporations are often the best choice for those seeking to finance their film. Corporate law is complex, so read the laws as they pertain to your state.
- When financing your film, don't forget grants and subsidies.
- When budgeting, the above-the-line costs fluctuate and the below-the-line costs are relatively stable.
- Stick with 35mm film stock.
- Be absolutely certain to have enough financing available to at least finish filming. Editing can wait, but unfinished filming is a total waste.

4

Actors and Agents

Before entering into this intricate and delicate field, there are a few remarks concerning players, professional and amateur actors.

Intrinsically the author has nothing against the use of amateur actors. Some of our finest professionals come from these ranks. I hired an amateur actor in a small role in a picture some time ago and regretted that he was a young yacht captain rather than a professional actor, because he could play in any picture anywhere. The low-budget director doesn't have the time to teach even the most basic rudiments of film acting while turning out a picture. He or she must perforce work with people who know what to do and understand just what is expected of them. Then, rather than teach basic principles, he or she can dedicate time to the finer points of directing a scene. Many moving and beautiful films have been made with amateurs, but those directors may have had more time, money, and patience.

One of the most publicized portions of the motion picture industry concerns the players. That is easy to understand because they are what the public sees, and they represent a large part of what the public pays for when buying tickets at the box office.

The element of box office appeal must be taken into account. There are good, bad, and indifferent players, but however bounteous the feast, only a small portion is available for consumption in a low-budget feature.

Motion pictures have practically no limitations; it is a question of using your imagination to the utmost. If you do have a choice as a director, there are many roads leading in all directions. The first and most obvious one leads to the agent. The yellow pages of major city telephone directories under "Agents" are cluttered with names, from the great international agencies to the obscure ones that deal in more questionable material. Somewhere between these extremes is the man or woman you are looking for, who can provide a remarkable package for your entire cast.

When good agents know your requirements, they usually can help you. Every agent has some ex-stars, aged or otherwise, on the way down, and you can usually get

Continuous Employment – Weekly Basis – Weekly Salary
One Week Minimum Employment

This AGREEMENT, made this ______ day of ____________ 19___ between ____ hereinafter called "Producer," and ____ hereinafter called "Player."

WITNESSETH:

1. **Photoplay, Role, Salary, and Guarantee.** Producer hereby engages Player to render services as such in the role of ________________________ in a photoplay, the working title of which is now ____________________________ at the salary of __________ ($ ______) per week. Player accepts such engagement upon the term herein specified. Producer guarantees that it will furnish Player not less than ____ weeks employment (if this blank is not filled in, the guarantee shall be one week).

2. **Term.** The term of employment hereunder shall begin on ________ on or about* ____________ and shall continue thereafter until the completion of photography and recordation of said role.

3. **Player's Address.** All notices which the Producer is required or may desire to give to the Player may be given either by mailing the same addressed to the Player at ____ or such notice may be given to the Player personally, either orally or in writing.

4. **Player's Telephone.** The Player must keep the Producer's casting office or the assistant director of said photoplay advised as to where the Player may be reached by telephone without unreasonable delay. The current telephone number of the Player is ________________.

5. **Furnishing of Wardrobe.** The (Producer) (Player) agrees to furnish all modern wardrobe and wearing apparel reasonably necessary for the portrayal of said role; it being agreed, however, that should so-called "character" or "period" costumes be required the Producer shall supply the same.

6. **Arbitration of Disputes.** Should any dispute or controversy arise between the parties hereto with reference to this contract, or the employment herein provided for, such dispute or controversy shall be be settled and determined by concillation and arbitration.

7. **Next Starting Date.** The starting date of Player's next engagement is ________________.

IN WITNESS WHEREOF, the parties have executed this agreement on the day and year first above written.

____________________ Producer	____________________ Player
____________________ By	____________________ Social Security Number

▶ ***Figure 7*** *Sample of a minimum free-lance contract. Note the "in or about" clause may only be used when the contract is delivered to the player at least seven days before the starting date.*

such talent at a small percentage of former fees. It is always wise to add at least one recognizable name to your cast.

Agents provide 8″ x 10″ glossy black and white stills, which tell you little about the players unless they are of fairly recent vintage. The height, weight, and complexion of your players is important because the cast must be balanced. All of us can remember the old B pictures where the hero and the villain were almost the same size, had the same hair color, and even looked alike. This process of balancing one character against another is usually an instinctive reaction, but should really be done on a conscious level, as it is of vital importance to the purely visual quality of the picture. Psychologically the villain, for example, may gain unwarranted sympathy because he is physically overpowered by a good-looking, muscular, and attractive hero.

To typecast or not to typecast — that is the question. The obvious advantages of typecasting are myriad. It makes for easy and fast directorial work because the ease with which the character is recognized and evaluated by the viewer moves the picture along faster. But the obvious is, or can be, trite and, unless counteracted by directorial ability, can be fatal to the reception of the picture. Usually low-budget films are typecast without question, although a few words should be said about casting against the type.

When a role is cast against type and the work is well done, it sometimes presents a more interesting picture and may be much talked about. It may become a hit because it is not trite or obvious. However, for an actor to play against type requires more thought and preparation as well as careful execution. The director must be constantly mindful of the dangers of the actor's slipping back into type and confusing the viewer as a result. Typecasting is a two-edged sword that must be used with caution.

Most producers and directors simply cast type. Another pertinent fact is that many actors are glad to try a different way after doing the same type over and over again. In some sensitive actors, however, this sometimes leads to desperate rebellion.

If possible, it's always good to see some film of the actors involved. Of course, if they are well-known this is superfluous; however, a video cassette or judicious viewing of parts of a previous picture will help you make up your mind in casting. Most agents can provide this material, although sometimes you will have to pay for the screening.

Contracts should always be treated with due respect by the production office. For your edification we are including a basic contract for this type of production (see Figures 7, 8, and 9).

Many actors belong to the Screen Actors Guild (SAG); thus the requirements are practically the same for everyone. Try to avoid trouble by staying within SAG guidelines. However, if you want to contract someone who isn't a member, go ahead and do so but be sure and treat them contractually just as if they were protected by the Guild. It's better to pay the going rate. Pictures have been made outside the Guild's jurisdiction. On examining these contracts, however, we usually find that the fees are the same as or above those stipulated as minimum. Making low-budget pictures does not mean that it is conscionable to be unfair to players.

A contract should contain all pertinent information and should be as complete as possible. When some specific act such as nudity, simulated sex, or work with animals, etc., is required, it should be so stated by scene number and explicit explanation. Have the players and agents read the contract before signing it. Leave no opening for "I didn't agree to do this or that." A contract with all details properly stated is binding for both parties.

One of the most touchy elements related to casting is the screen credits. These should be established from the beginning. If you are using a new male star with an actress with some experience backed up by a once very well-known star of fame, you are presented with a situation which, unless dealt with at the start, could present difficulties later on. The solution, "Jane Doe," followed by "John Doe," and special guest appearance of "Larry Moe," might satisfy everybody. The enormous crawl at the end of the picture has little importance because the public generally doesn't stay to read it and most projectionists cut it off anyway.

Be sure to include in the contract the start and finish dates of their work. Players take great interest in them because, even if they don't have an upcoming engagement, they may have one before the picture is finished and it is only fair to them to have this security. Also the manner and times of remuneration should be explicit. No one likes to wait for payment of money earned. Additionally, when you talk initially with the agent, make it known that there will be no percentage deals and no deferments.

Employment of Day Player

Company ______________________
Date Employment Starts __________
Part ___________________________
Production Title ________________
____
Production Number ______________
Daily Rate ______________________
Weekly Conversion Rate __________
Married ________________________
Date of Birth ____________________

Date ___________________________
Name ___________________________
Address _________________________
____
Telephone Number
____
Social Security Number
____
Legal Res. of What State _________
Citizen of U.S. __________________
Quota Number ___________________
Date of Entry U.S. _______________

____
Producer

____
Player

____
By

▶ *Figure 8 Sample of a day player employment agreement.*

Authorization to Reproduce Physical Likeness Release

For good and valuable consideration, the receipt of which from is ____________ ____________________________ acknowledged, I hereby expressly grant to said ____ and to its employees, agents, and assigns, the right to photograph me and use my picture, silhouette, and other reproductions of my physical likeness (as the same may appear in any still camera photograph and/or motion picture film), in and in connection with the exhibition, theatrically, on television or otherwise, of any motion picture in which the same may be used or incorporated, and also in the advertising, exploiting and/or publicizing of any such motion picture, but not limited to television or theatrical motion pictures. I further give the said company the right to reproduce in any manner whatsoever any recordation made by said company of my voice and all instrumental, musical, or other sound effects produced by me.

I hereby certify and represent that I have read the foregoing and fully understand the meaning and effect thereof and, I have hereunto set my hand this ______________ day of ____________________, 19___.

Witness: ____________________________

____________________________ ____________________________

▶ ***Figure 9*** *Sample of an authorization to reproduce physical likeness release.*

Percentage deals lead to accounting difficulties ending in audits and, sometimes, in expensive lawsuits.

If there are potentially dangerous actions in the script, these must be mentioned in the contract with the stipulation of doubles for said action. If the action is not overly difficult or dangerous and the player agrees to do the scene, it should be so stated. It is better to be safe than sorry.

The producer's obligations regarding transportation, travel time, hotel accommodations, meals, wardrobe, makeup, and per-diems should be accurately and amply stated in the contract.

Open-end clauses are always dangerous. Many years ago in a big production, a star was granted a Rolls Royce and chauffeur for the duration of a picture. For many inexplicable reasons the shooting dragged on sporadically for more than a year. This can give you an idea of what can happen when contractual clauses are not properly thought through to their logical conclusions. When in doubt, have your legal counsel read all contracts before presenting them for signature. You may say that many of the things mentioned in this book are obvious; indeed they are, but many of the obvious matters are overlooked or simply forgotten in the headlong rush to get a picture filmed. Never underrate the obvious.

In considering the financial aspects of casting, remember one outstanding dictum: fit the cast to the available budget. Find out exactly how much money is on hand and look for players who will work within that sum. Do not expand the budget for this or that actor. If you go over the budget, you may finish the picture, but it may not recuperate its total cost in sales and rentals. The profit margin on this type of feature is not large enough to take unnecessary financial risk.

If there are enough finances available for a better cast, that is wonderful: the picture should be in the upper bracket of independent production. Good direction and camerawork can make a player perform better than usual, and that depends not on finances, but on talent.

SUMMARY

- For a low-budget feature, hire professional actors. Spending the time and money involved in teaching someone the ropes should be left to those who have plenty of both.
- Work with an agent who is comfortable with and has experience in working for low-budget productions.
- Fit the cast to the financing available.
- Add at least one recognizable name to your production.
- When casting against type, be certain that you and the actor are up to the challenge. It can be rewarding but, when done badly, is disastrous.
- When casting type, don't be trite or obvious.
- Take great care with the contracts. Don't leave any open ends and be sure to specify exactly what you will and will not do.

5

Preproduction Preparations

Scouting locations for filming represents quite a list of map studying, driving, flying, and generally moving from one place to another if your script calls for varied locations.

"A rock is a rock, a tree is a tree — shoot it in Griffith Park," as an oldtime Hollywood producer once said, hardly applies anymore; however, the writer is quick to state that the first great picture of a now-famous cult director was shot in Griffith Park not too many years ago, using artificial snow and inexpensive sets, rather than in the striking beauty of the Sierra Nevadas, because of the insistence of the much-belittled producer. Probably the picture was better for it because the director was forced to use his inner discipline and concentrate on the psychological aspects of his drama, utilizing his excellent actors to the maximum of their abilities, rather than photographing the grand, austere beauty of the other location which would have deprived his cast of the proper visual attention.

The list in Chapter 2 on breakdowns should be consulted and assiduously discussed before visiting the location. Accessibility and transportation to and from the site with corresponding travel times must be taken into consideration in all cases, even when the location is nearby.

On these scouting trips, use discretion; otherwise everyone will think a multimillion dollar film project is coming to town and will sharpen their teeth in expectation.

Once the locations are listed, they must be visited with camera in hand. Some directors prefer to use a video cassette camera for filming on-location trips, but usually a 35mm still camera is sufficient, accompanied by a well-filled notebook. One-hour film developers provide quick, efficient service, and you will have a picture to hand to the producer if he or she hasn't accompanied you on the journey. This is an economical way of comparing locations, and is akin to much more expensive story-boarding. The locations chosen should have visually interesting features that serve to enhance the meaning or intent of the script.

If these trips were used solely for the above-mentioned purpose, they would be expensive; however, one stone can be used to kill several birds. Housing in local motels and hotels is considered in-depth. For most purposes a good, clean, reasonably priced hotel should be located close by. At this time the production office has provided an approximate shooting schedule and cast list, which will permit you to calculate the number of rooms needed.

On a low-budget picture, the crew members usually double up on rooms; exception is the cameraman, who rates a separate one, just like a star.

At the same time that you are looking for hotels, study the catering situation. With one hour for lunch, travel time becomes important. If the location is five to ten minutes away, then it's possible to contract the midday meal at a restaurant, getting a price break for quantity. Fifteen lunches for 24 days is 360 meals! If distance makes this type of meal impossible, then a delivery catering service must be utilized. In and around motion picture centers there are always caterers who understand the needs of motion picture producers and who are eager to offer their services. Their prices and what they offer must be checked against similar services.

If none of these services is available or the location is distant, it will be necessary to hire a person to prepare the meals on the site. As any sergeant knows, soldiers will stand quietly for almost anything, but will protest about bad food. It's the same with film crews. Just be sure it's good and enough.

When the start dates and contracts are signed, go right ahead with reservations and deposits if this location is what you are looking for. By doing this in advance, you are sure to have all your crew together where you can keep an eye on them.

Make one thing clear to the management of the hotel: Any telephone calls and additional services are not the responsibility of the production company, and will not be paid for by said company unless stipulated by contract. Put it in writing, because the additional expenses can be staggering.

Nowadays most states, as well as some cities, have a film production office. By all means, take advantage of it for a number of reasons. Some offer books and booklets with photographs, addresses, and a commercial directory of the area, but, first and foremost, they can facilitate permits for filming almost anywhere in their jurisdiction. Some states charge location fees for what used to be free. By consulting the film commission office or board, this onerous paperwork can be reduced and savings made.

These offices can also facilitate the use of security guards, off-duty policemen, or state troopers. Their salary equivalents must be paid, of course, but in some areas their aid is almost indispensable. They will control traffic, set up sidewalk and road blocks, and generally supervise onlookers who can sometimes cause interruptions.

Once in a while, officials of small towns like to demonstrate their authority, especially to out-of-town film folks. A quick call from the Governor's office usually puts the train back on the track in a hurry.

On one production, a very small-town pro-tem mayor, who probably felt slighted for some obscure reason, served the company with a cease and desist order (which was promptly ignored because the filming was on private property). This type of thing really does happen, so be prepared in advance.

One of the most useful public entities in existence is the local fire department. The members, mostly volunteers, are usually moved by a strong sense of civic duty.

For a modest fee or contribution they will be very helpful. Most small town units don't have much opportunity to show off their abilities or even to practice, so working with a film company gives them a unique possibility for public display with the consequential publicity.

On one low-budget production, a magnificent storm, day-for-night scene was set up with the aid of a volunteer fire department and 8000 gallons of water. A helicopter was standing by for wind effects, but wasn't really necessary as a strong wind was blowing in from the Gulf of Mexico onto the dunes where the production was filming. With all the movements and supply of water from the fire engines controlled by walkie-talkies, the entire complex scene was in the can in less than three hours, before the lunch break. A fireman is a good friend to have in time of need.

Over the years the Army, Navy, Air Force, Marines, and Coast Guard have aided innumerable productions by appearing briefly in or throughout entire pictures. Today the armed forces are still wiling to help out, but are much more discriminating in their choice of films and the extent to which they will participate. The principal reason is the cost factor. All of the armed forces feel, and rightly so, that they should not appear in productions that are defaming their good names or derogatory in any sense.

If their intervention in a picture is small, permission can usually be obtained from a local commander. If their participation is on a larger scale, permission must be obtained from the corresponding department in Washington, D.C. For low-budget films, let us assume the former rather than the latter situation prevails. Do not forget to give military cooperation appropriate screen credit for their efforts. They do appreciate and deserve it.

One of the most vital points of preproduction planning involves the equipment to be utilized. Many persons just entering or planning to enter the motion picture field are obsessed with the idea of purchasing equipment. Simply owning it makes the whole project closer, more real, and definitely within the realm of possibility. The writer knows, because at some time or other he has owned practically every piece of filming equipment imaginable, both 16mm and 35mm, even up to a small stage. Take the word of one who has been there: Owning all this equipment is neither practical nor economical. Equipment that isn't working incurs losses. The same money in a savings and loan institution or any other investment would produce income to help you wait out those dry periods we all have.

Rental equipment is cheap, economical, and accessible, anywhere, anytime. If you want equipment in Timbuktu, there's a company that will fly it in for you to use tomorrow. In the United States, Canada, Mexico, Venezuela, Japan, Argentina, Great Britain, France, Germany, Italy, Spain, Hong Kong, Australia, and probably some places that we've never heard of, there are companies that can supply what you need, literally overnight.

Equipment rental companies are everywhere, but some are better than others. The large Hollywood or London companies may offer the best deals. They are geared to providing for the most exotic of demands, as well as supplying garden variety equipment; they are also willing to give competent advice should you need it. If you

LIST OF RENTAL EQUIPMENT

1. Two Arriflex IIC cameras
2. Four 400′ magazines
3. Four battery belts
4. Two matte boxes and lens shade
5. Two highhats
6. One Angenieux 25-250 zoom lens
7. Two sets Schneider lenses (28, 50, and 75mm)
8. Gelatin filters (85, 85N3, 85N6)
9. Two variable speed motors
10. One standard tripod
11. One baby tripod
12. Two O'Connor 100 fluid heads
13. Two triangles
14. One Luna-Pro exposure meter
15. One Gossen Sixticolor meter
16. One changing bag
17. One slate
18. One stopwatch (16-35mm footage)
19. One tape measure
20. One Nagra tape recorder
21. One directional microphone with cable
22. One fish pole mike boom
23. One mike stand
24. Five 1000w colortran focusing lamps with stands
25 Two 1000w colortran Broads with stands
26. Two 1000W Lowell mini lights
27. One colortran portable 30v mini-pro with battery
28. Four 4′ X 4′ sun reflectors
29. One super grip car mount
30. Two polecats
31. Assorted cables, boxes, and switches
32. One moviola — Editor's model
33. One synchronizer — 4 or 6 gang model
34. Rewinds, reel horse, reels, flanges, and cores
35. Bins, bag tables
36. One CIRO tape splicer
37. One 35mm portable projector
38. Two ladders (one short, one tall)
39. Four location umbrellas
40. Six folding director's chairs
41. One portable projector screen

Many other items, small or large, may be required on any given production. The above list serves as a guide to minimum requirements. All filtering material such as Rosco media must be purchased separately. One complete first-aid kit should be available at all times.

pay promptly and make your up-front deposit, you'll find they will meet you more than half way. A cautionary note at this point — some companies make outrageous credit-check demands, etc. You will find, as in most things, the best rental houses give the least trouble.

The first step is to prepare an equipment list. We have included one for your perusal. You will notice that it is not very long, but it is sufficient to shoot a full-length feature without too much trouble. Don't be misled by advertisements recommending some piece of special equipment that you can't live without. When we look at the modest equipment used in years past to film great pictures, we might well be ashamed at the extent of our own list.

At the top of the list we have noted two Arriflex IIC cameras with their corresponding motors and magazines. Why two?

On even the most modest production, there will be times that the director needs two cameras to cover the action, but the real reason is to provide a backup. If, for any reason whatsoever, you lose or damage a camera or it malfunctions, the other one is working in a matter of minutes. The same goes for battery packs or belts. Four magazines should be ordered and kept loaded at all times. These, by the way, should be the 400-foot ones ordinarily used. A set of Schneider lenses (28mm, 50mm, and 75mm) should also be ordered. If the script calls for any special wide-angle lenses, add a 21mm or 18mm to the list. Rarely will you need anything wider than these.

An important item on the lens list is the 25-250mm 3.8 Angenieux zoom lens, or its more modern equivalent. We will go into detail of its use in Chapter 10 on Directing the Cameraman.

The filters are one of the most important items on your equipment list. First are the color compensating filters; nowadays only one of these is necessary — the 85. For the standard lenses, glass filters are available, but the gelatin ones are usually cut from a square and mounted behind the 25-250 zoom. These filters are not automatically supplied and must be ordered specially. There are also available a press-on holder and circle cutter, which are very handy and should be ordered. It's also wise to have several neutral-density filter combinations.

The second category of filters is for color temperature control of doors and windows, and comes in 50-foot rolls. Rosco, Inc. has samples available of all the working combinations. Nothing is quite as distracting as a blue-white blast of sunlight from an unfiltered window at 5600K, combined with incandescent lighting at 3200K: they don't mix very well, despite what some technicians say. These things distract attention from the actors, subliminally diluting their performance.

Figure out what you need of 85N3 or 85N6 and order it from your rental house. This must be purchased as it is not a rental item. Another indispensable filter material that should be ordered is Daylight Tough Blue 50, which comes in 25-foot rolls and is used to balance your tungsten or quartz lights when they are used outdoors or in shade.

Be sure to get two fluid heads for the tripods. There are many available, but the O'Connor 100 has always proved satisfactory. Friction heads make life difficult for the camera operator and lead to juddering pans and tilts if not handled properly. With Arriflex cameras, don't forget the hi-hats.

A standard tall tripod, as well as a baby one, comes into use practically every day. Be sure to get a couple of triangles to put under them.

A changing bag is essential, though often overlooked in the last minute rush. Make sure you have one.

Most cameramen prefer to use their own light as well as color temperature meters, but it is a good idea to back these up with the rental items.

One good suction mounting is worth its weight in gold. It's difficult to remember what we did before we had them.

At this time, you should make your final decision on lighting equipment. The list of 10,000 watts of Colortran Quartz lighting equipment should more than fill the bill. You'll notice several 1000W Lowell Lites, which are extremely useful for close quarters. If you are going to take interiors of a moving car and need fill light, be sure to order them. Dull, available light shots of the driver and passenger look poor. Use that fill light to avoid bland car interiors. Be sure to list the stands, polecats, clamps, and gaffer tape that you'll need, and get enough cables, distribution boxes, switch boxes, and fuses to take care of the lot.

We should mention sun reflectors here. This is a thorny question. Our advice is: Don't rent them; make them. Buy two sheets of 3/8″ x 4′x8′ three-ply, and cut them in two 4′x4′ panels each. Use white glue to stick on aluminum kitchen foil, being sure to wrinkle it well and then smooth it out before gluing it on the board. For just a few dollars your production will have four sun reflectors and no rental or transportation problems.

If a large amount of fill light is needed for outdoor work, it may be necessary to rent one or two Minibrutes. These nine-lamp units not only provide an enormous amount of light, but also draw plenty of current. Be sure you have ordered enough heavy duty cable for these, because ordinary ones won't carry the load. Since these must tie into the 220-volt lines, have an additional heavy-duty fuse block and throw switch on hand. Minibrutes require no generator or operator and are easily handled by the regular crew. For some pictures they are indispensable.

You must have an AC-DC volt-ohm meter to make your gaffer happy and keep him or her from hooking up 110-volt lights to the 220-volt line and blowing up the works.

The sound recorder should be an ordinary Nagra with a fairly directional microphone. Get a simple playback unit to review the sound. Sophisticated shotgun mikes or very expensive late-model recorders are not necessary for this type of production. Get cheap and reliable standbys, just as you would cameras. We will go further into sound techniques in Chapter 12.

In addition to all this equipment, there are many useful little items that you'll need every day, such as a toolbox containing a hammer, nails, pliers, cutters, screwdrivers, empty plastic hand-operated spray bottles, etc. Try to assemble all these items beforehand to avoid emptying the piggy bank of petty cash.

Film purchase, raw stock, negative, whatever you want to call it — the stuff that runs through the cameras at 90 feet per minute, the purveyor of that vital image which represents the end product of all this furor, hassle, and insomnia — is a vital concern. There are ways to save money on this item but, in this case, you do so at your own risk. There are several places where you can buy leftover, unopened cans of film, short ends, etc., at very reasonable prices. However you should understand that manufacturers of negative, and all film material, in fact, even the best of them, make changes in their products from batch to batch. Normally these are subtle changes in gelatin ripening, chemical quality, etc., and so each lot is given a very important batch number. But sometimes these changes are not so subtle and if materials from two different batches are used, the resulting products will look extremely different. Another factor involved is aging. Negative material, unlike wine, does *not* get better with age. Part of the unused stock may have been in the tropics and another part in the desert or in cold storage.

Let's say you are running two cameras on a fight scene where two men are having it out, and the shots from these two cameras must intercut in a fast sequence. If you happen to have different material or batch number on each camera, the resulting jumps in color and density will make both you and the timer at the laboratory quite unhappy. Trying to even up the sequence will drive either or both of you to drink or the psychiatrist. True, you can save some money by buying this kind of film, but is it worth it? The alternative is easy and, in the writer's opinion, the better. Go to your supplier and order all the negative at the same time. Calculate how much you'll need and try to get all of the same batch number if possible. If you are in constant production (happy thought), most manufactures will set aside entire lots for your exclusive use.

Be sure to store your film in a cool, dry place, but do not refrigerate it unless you're working in the Sahara Desert. It takes too long to bring it up to room temperature. All the magazines should be loaded the night before shooting; be sure to keep a white cloth or a large umbrella if you have one over the camera when it's not in use. A final thought: number your magazines and label everything.

While on this track of celluloid, let's get to an all-important question: choosing a laboratory. Technically speaking, most labs are competent and can do a good job. The differences reside principally in the small additional details of service that make life bearable. Oddly enough the size of the lab seems to have little bearing on this; we have used both small and large labs with the same results. On the whole, it is better to pick a lab that handles feature films on a daily basis. They are well aware of many of your problems and will do their best to help you out.

Labs will supply their favorite form for camera reports and stick labels for your cans. Fill these out punctually and as fully as possible, stick them on, and see that the copies of the reports get to the right people. A case comes to mind of a prominent director, well-known for his irate character, who had the misfortune of losing a 1000-foot roll of vital negative. The results were marvelous to behold. By reshooting some tight close-ups and using a few sundry editing tricks, he finally made his way around the loss.

All lab catalogs contain many pages of prices for different work that they do, and many more pages of disclaimers of responsibility. Use the stated figures only as a preliminary guide to costs and ignore the disclaimers. It's bad for your ulcers to read such printed matter. Instead, go to the lab representative who handles the feature films and have a heart-to-heart talk. The lab rep is your best friend and don't forget it.

He or she will ask the name of your feature, when you intend to shoot, how many feet you are going to use, how much work print you'll want, and so forth. With all the pertinent information in hand and the help of an electronic desk calculator, the lab rep can come up with a figure up to and including the answer print, that will make your producer very happy. The overall price will be much less than you expected. At the same time you should receive a contract and credit application. Make your payments promptly and get a good name in the industry, which will make the process easier next time.

If your financial situation is not too clear or your vendors' confidence in you is a little shaky, they can become difficult. Remember, they hold a lien on your negative, the details of which must be satisfied before release of your negative. Most labs will be happy when you tell them that you intend to make all your prints with them, but be careful not to compromise yourself on how many copies or when you are going to make them. The subject of copies will be addressed in Chapter 17.

Now, this material must be taken to the lab and the rushes or work print have to go where they are needed. First consult the lab about their preferred means of transportation. Be sure to pack the film properly. If you can get plastic foam containers, all the better. Be sure to ship the exposed negative as soon as possible so that any defects or problems will turn up right away, allowing you to reshoot before leaving the scene of the work. This is the ideal. UPS, overnight air freight, etc., can save you a lot of time and wear and tear on your nerves.

Sometimes it's necessary to cross international borders on location filming, which means going through customs. Every country has its own peculiarities that must be taken into account before making any final decision. Investigate the permit requirements, supervision, censorship, and idiosyncrasies involved, and be sure someone from your organization speaks the language of the host country, understands it very well, and can deal with the problems involved. We personally have crossed borders in filming a low-budget production and affirm that it can be done, but it is not always the easiest thing to do.

There is also the question of visas to abe taken into consideration. Visa requirements for foreign technicians working in the United States have changed recently, making it difficult to employ them. Conversely, as a result, visa requirements of the countries involved have also stiffened. These things change from time to time, but you should investigate them thoroughly before production starts.

Here is one concrete example. By federal law, the Mexican feature film industry is a dictatorial closed-shop, controlled by six interlocking unions, with whom six separate obligatory contracts must be signed before beginning production. Filming features is prohibited outside this context. The technicians' union, for example, stipulates that a crew of 41 members must be hired, regardless of the size or scope of the picture. At the same time, the directors' union requires a standby Mexican director from their union to be hired at full salary, whether he is needed or not. This is only two of the six unions involved; writers, musicians, actors, and composers follow. Finally, the Secretary of Interior requires full censorship of all material filmed, and the censor's fees must be paid for by the production.

Our neighbor to the north, Canada, also has many rules and regulations and the situation in Europe is similar. All savings accomplished by lower laboratory costs and salaries are obliterated by the staggering requirements that can fall on you the minute you step across a border, since these dicta involve a whole country.

Another example comes to mind regarding a filming in the Galapagos Islands off the coast of Ecuador. The film company found that the red tape easily reached the moon or beyond. The parties involved will naturally deny these allegations as being without foundation, claiming they are simple regulations to maintain order and so forth, but the truth is as stated. To beat the game, you must know all the rules. The obvious answer is to hire a production supervisor experienced in the chosen country, but in general, avoid crossing international borders when you can. Crossing state lines is bad enough.

While shooting on location, certain comforts and necessities must be taken into account. Hollywood studios have specially prepared units for all the requirements. Some rental companies have portable dressing rooms and other such conveniences. However, these choice items make large holes in a low-budget producer's pocket. How can this needless expenditure be avoided? Easy. Rent a large recreational vehicle with refrigerator, shower, toilet, stove, etc. Hire a competent driver to take it to and from the location, attend it, and keep it clean and serviceable. Stock it with supplies for morning and afternoon coffee breaks. Forbid all liquor on board, including beer, and be sure the RV is used for legitimate purposes and not for loafing. Alcoholic drink of any kind does not mix with motion picture production. If your crew or players want to drink, let them do it with their own money and on their own time.

The 11 A.M. and 4 P.M. breaks are traditional and necessary. Coffee, tea, or a cup of instant consomme with a cookie, cracker, or light snack makes work livable. Of course, the director usually continues working with a coffee cup in one hand and the breakdown in the other, with a harassed script supervisor trying to get a quick cup and still come up with an answer for "What was he wearing in take 268?"

It is advisable to rent cars where necessary, rather than use personal vehicles, which entails additional insurance problems. A 12-passenger van is very useful to transport personnel and equipment for this type of production. It also saves on the bookkeeping entailed with the use of a dozen different vehicles.

All airline tickets should be purchased and reservations promptly made to avoid logistical hangups later on; certain vacation periods may interfere with your production schedule, but usually this can be avoided, not so much by shifting your schedule as by making the reservations on time.

All of the preceding chapters may sound like filmmaking entails a lot of paperwork — it does. This is all part of the procedures which will allow you to beat the game. It's a well-known fact that a month's work at the production office costs less than one day's work on the set. By careful preplanning, it is possible to calculate, within reason, the needs of all the personnel and the equipment necessary for the entire production. In case you are a little off in your calculations, there is a final item on all budgets — the already mentioned contingency figure. It's a good cushion to have, but try to save every bit of it. You may need it all before you finish.

SUMMARY

- When scouting for locations, also scout for motel accommodations, catering, and transportation.
- Be sure to know what the local regulations and fees are before committing to a location. See if there is a local film commission. This is especially important when filming abroad.
- Rent your equipment. Renting is economical and accessible.
- Prepare a very thorough equipment list.
- Find a reputable lab and become their best customer. They may save your film before you are through.
- When crossing international borders, know your host country's regulations and peculiarities.

6

Hiring the Crew

The crew members are your soldiers. For better or for worse, you are going into battle with them for a month, more or less. To form your working group, try to get seasoned professionals whenever possible, for they have been there before and most of them will obey any reasonable command.

At this point a delicate matter must be taken into consideration: the relation of the low-budget producer to the union organizations involved in the industry. Most of the producers in this category cannot afford to hire a full union crew. The reason is not that the producer is against the unions, but that most of them are geared to the large productions of the majors and their affiliates. As the unions cannot guarantee full-time employment to all their members, many of these work on low-budget productions. Their salaries may be the same, but the crews are smaller.

The complete crew and staff will consist of the following: director, assistant director, assistant producer, script supervisor, sound technician, "gopher," cameraman, assistant camera operator, gaffer, assistant gaffer, and two helpers — 12 persons in all. It is entirely possible to work with less personnel; one production of a full-length feature comes to mind that was done with a crew of six and two helpers, but there is such a thing as overwork.

Let us assume that the production is non-union and is being carried out in a right-to-work state. There are about 19 states where union pressure cannot legally be used against your company to force you to hire union personnel.

As a producer, you should know how many technicians are needed to do a given job. An easy low-budget picture can be done with one cameraman with assistant, one gaffer with assistant, and two helpers on the crew. This is a basic crew of six. If your picture is complicated and requires more locations, actors, or action sequences, a larger crew will be necessary.

I would stress one point about the personnel: they should be professionals with feature film experience. This is the old catch again: you can't work if you don't have experience, and how can you get it if you don't work; but here we are considering the best and most economical way to get a picture made, not how to resolve this dilemma.

Your helpers can be anyone, manual workers or persons interested in working in low-budget pictures, but the technical crew must be professional.

An assistant director who is lively and competent is well worth his or her salary, because this person will take care of all the minor details on the set before the director starts shooting the sequence. An assistant producer who doubles as bookkeeper in this kind of production is essential. The script supervisor handles continuity and reports. The sound technician is a necessity in a low-budget picture because he or she records the direct sound that must be used as a guide track. A stills photographer is needed for publicity that has to be made in every production, but if you don't have somebody to do it, there is an alternate solution. Every feature film needs about 35 to 50 good still photographs, usually in black and white, but some choice ones are needed in color for the poster, sometimes called the "one sheet," the famous press book, and general publicity.

A director who has been working 18 hours a day has no time to worry about this problem. The cameraman and his assistant are hard-pressed for time, so the stills simply don't get made. One option is to wait until the picture is finished and then make the stills. Here is how: Mount a 35 SLR camera with zoom lens for copying transparencies on a baseboard, and set up a light source in front of it. This light can be a flash or photo floodlamp. Set up a rewind and reel horse on the baseboard and, using the answer print or any other copy for that matter, start shooting all the frames that you would like to have from the picture. You'll finish up with several hundred black and white negatives. Prepare proof sheets and select the best scenes. Send off the chosen few to be multiple-printed at labs that do this work for a very modest charge, and you have your stills. Color presents a little more work due to contrast buildup, but this can be solved by using appropriate film and development. This is the easy and low-cost solution to the absence of a stills photographer.

The cameraman should be the camera operator. Eschew the director of photography, who is not absolutely necessary in a low-budget picture. The camera operator must know how to do the setups and do them fast. The assistant cameraman has to know how to load magazines and change them fast, as well as pull focus. On big productions, cameramen have a special focus puller, but this is an extra expense. The gaffer must set up the lights without undue delay. Your director has to know right away where the action is, because there is no time or money for fooling around.

A good cameraman is a combination of varied qualities and experience. One of the most important factors consists in his personality. Good camera operators have an instinctive feeling for lighting and can interpret the qualities of the screenplay in this respect. Their orders to the gaffer must be precise as to their requirements. The cameraman is the "First Mate" between the director and the crew and, if he really knows his craft, will be in command of subordinate personnel. Panning, tilting, and dollying to obtain the initial and final framing prescribed by the director is an art and is based on the sensitivity, muscular reaction of the camera operator.

Juddering pans, sloppy tilts, and indecisive tracking shots have no place in professionally made films. Like any good first mate, he will obey commands and, when the situation demands it, make pertinent suggestions to the director. Where are we going to find this paragon of sobriety and perfection? Good cameramen are always willing to demonstrate their work, but avoid a cameraman who is a frustrated director and would rather do your job as director than his.

Usually the cameraman will have a favorite assistant who remains with him from picture to picture. The two should work together very well as a team. The assistant is one of the most valuable elements of the production, anticipating the cameraman's desires and having everything ready as it is needed. The assistant is in charge of magazines, battery belts, meters, tapes, chalks, etc., a jack-of-all-trades with a lot of little tools and oddments which come in very handy in this kind of production. The cameraman depends on the assistant just as the director depends on the cameraman, who will say to the producer, "I want so and so as my assistant — we work well together."

Every permanent person on your crew must have a contract stipulating their obligations, emoluments, when and where they have to work, and so forth. This contract must state the name of the picture, where it is to be shot, how long it will take, and overrun agreements, as well as overtime pay.

Contracts! Contracts! Will this paperwork never end?

No, not even after the picture is completely finished!

It all must be there. Personal relationships don't matter, legality is legality. The final picture may be art when it's finished, but until then it's strictly business. So you are now a professional engaged in writing, producing, directing, filming, acting, or helping make a full-length feature film with all the obligations and requirements pertaining thereto.

The gaffers on a low-budget feature film are persons of many talents. Around their belts hang cutters, pliers, screwdrivers, tape, and many other things that you would never think of. On the set the gaffer must know where to tap into 220-volt mains with unerring certainty. He or she knows how much the fuses will carry without resorting to calculators, tables, or consultation, and when the lamps are running low or are too hot. The speed and rhythm of your workday depends a great deal on this person for the cameraman can't say to the director, "I am ready when you are, J.R.," until the set is properly lit.

We have deliberately left for the last that most important person, that indispensable part of your staff, the overworked slave in a low-budget production, the script supervisor. The script supervisor takes camera running times and makes all miscellaneous notes pertaining to the scene that is being shot, writing them in the script. Although not absolutely necessary, to make the job easier, script supervisors should be able to type. Sometimes they may even have to make out camera reports and keep a spare copy for the director. They follow the wardrobe list and consult the prop list from time to time, write reports for the producer of the day's work, and note such things as "take 456 - 5, print take 2 — hand on red tie, place cup on table. 10:58." "Be sure to have empty bottle on table. Call hairdresser appt. Miss Jones," "Print take 2 - 23." Script supervisors catch mistakes before they are committed forever to film, and represent a most important link to the production office. They also carry a notebook, a breakdown, and just about everything else that is required. The director depends greatly upon the script supervisor.

Now for a word on helpers of a low-budget picture. It seems strange, but when you want to hire them you can never find any. Thousands of students of film would love to be helpers but can't leave all their classes for a month. You read and hear of aspirants who would carry a stand, sweep the floor, or do anything just to work on movies — but where are they when you need them? On a recent production the State Employment Office was called, among other sources. Of course we found helpers in the end, but it was difficult. But never mind my petty carpings; we were making the movie, and so far it was within budget and on schedule.

While you were reading about the pitfalls to be avoided on the way to the answer print, please don't think I was trying to dissuade you from making films by stating plainly all the negative aspects. On the contrary, I am trying to explain the process the best I can so you will be able to make your picture. If you avoid even one of these negative aspects, the purpose of this dissertation will be well-served.

SUMMARY

- Hire seasoned professionals. They have been there before and their experience will prove invaluable.
- A complete crew will consist of a director, assistant director, assistant producer, script supervisor, sound person, "gopher," cameraman, assistant cameraman, gaffer, assistant gaffer, and two helpers — (12).
- On many low-budget, nonunion films, you can get away with a smaller crew consisting of a cameraman, assistant cameraman, gaffer, assistant gaffer, and two helpers — (6).
- There are 19 "right-to-work" states where union pressure cannot legally be used to force you to hire union personnel. However, do not use "nonunion" as an excuse for poor pay.

7

Putting the Production Together

Every picture has to be filmed somewhere. In a previous chapter we mentioned a list of locations. Now is the time to put this list to complete use. As many rental arrangements as possible should be made in the preproduction period, to lighten the workload during filming. If a sequence demands the use of a yacht harbor, for instance, visit several to see which one corresponds to the mood of the script. After locating it, make a film commitment as to price and dates. Some bargaining usually has to be done, as most managers hear the word "picture" and Hollywood's millions show up in their eyes. Be truthful from the beginning; state that it is a low-budget film. If the price is not reasonable, look for another place.

Studio rentals are prohibitive except in certain circumstances. Smaller sound stages are available, but everything must be taken into account in a financial feasibility study, i.e., can we afford it or not? Remember that in the Hollywood area there are companies that can provide flats, windows, doors, and all the apparatus to put together a set very quickly at reasonable cost. Investigate all these possibilities. Pictures have been shot in vacant warehouses using this method.

When interiors are needed, such as apartments, offices, houses, etc., your best bet is to visit real estate agents specializing in furnished rentals. In most cases they are willing to cooperate with motion picture companies, although some of them have been badly burned in the past by misuse of their property. There is a story, probably apocryphal, of a prominent person's recognizing his estate in a porno movie after having rented it for the summer to a respectable old lady from Pasadena!

Many of these rentals are well-furnished, requiring only minor changes to satisfy the demands of the picture. Finally, they come cheap compared to putting together a set. The same applies to drive-ins, motels, hotels, and other locations.

A pertinent fact to be taken into account is that some states, like California, charge high fees for daily use of state beaches, parks, mountains, etc. When you are not sure where your state line is, talk to a friendly sheriff who may be able to help you.

If your picture is a Western, remember the old saying, "When in doubt, make a Western." You are in luck. There are western towns available in Southern California,

Arizona, New Mexico, Texas, and Utah. With some judicious changes in decor, you have a different looking town. There are rental outfits to supply all the western gear you need, from horses to handguns, stage coaches to branding irons. Costume rental companies can dress your entire cast with astonishing speed. If your director doesn't demand that you import a vintage steam locomotive and train from some museum, you will be pleasantly surprised at your budget requirements.

There is one minor but quite necessary piece of paperwork to be carried out with these locations regarding rental transactions. It's called a property release (see Figure 10). You'll need a few of these, so stock up, just as you already have on releases for recognizable persons who appear in your picture; without these signed

Property Release

To: ______________________________ Date: ____________________

____

____

For valuable consideration receive, you hereby grant to the undersigned permission to go upon your property located at: ______________________________
______________________________ for the purpose of photographing certain scenes in connection with a motion picture to be produced and undersigned, tentatively entitled ______________________________.

You agree that we may have the right of use, possession, and access to said property as may be necessary for such purposes.

We agree to use reasonable care to prevent damage to your property and to remove any and all property which we may have placed upon your premises in connection with our use thereof. We agree to restore said property as nearly as possible to its original condition at the time of our taking possession thereof, reasonable use thereof excepted. You hereby give and grant to us the right to photograph said property or any part thereof, and to exhibit and distribute photographs thereof in connection with any such photoplay or in connection with the exploitation and advertising thereof.

For this right of use of your property we agree to pay you as follows: ____________
____

We hereby agree to indemnify you and hold you harmless from any claims and demands of any person or persons arising out of or based upon personal injuries and/or death suffered by such person or persons, resulting directly from any act of negligence on our part while we are engaged in the photographing of said photoplay upon your premises.

Starting Date On or About ____________________

Approved and Accepted
By ____________________ ______________________________
Producing Company

____
By Authorized Agent

▶ ***Figure 10*** *Sample page of a property release agreement* releases you can face a lawsuit. Of course, this does not apply to shooting a public street open to anyone; it applies to the use of specific properties. It takes only a couple of minutes. When the paperwork is being done, just hand the release document over to the interested parties for signature and tell them, "It's a mere formality; thank you."

During the last days before the start of the picture, you must visit your insurance agent, truly an important figure in your life as a producer. Insurance is not cheap, but it's invaluable.

The first item to be handled is worker's compensation or whatever name it's known by in the state where you're working. Since you have all the names of the personnel involved, this will be easy to do. Some states charge premiums for the entire year, but under any circumstance it must be paid, so get on with it. Your accountant can help you by listing state and federal requirements. By all means, comply with everything to avoid bureaucratic hassles later on.

Then comes your general coverage for damages. This depends a great deal on what you are using in your picture. Be sure it's enough coverage, but no more than the sum required to replace that yacht if you inadvertently sink it! A blanket coverage of all actors and personnel is sufficient. Do not forget the clauses covering loss, fire, and theft. All of these things can and do happen. If you are covered for all these accidents, you are safe or relatively safe.

Low budgets rarely have enough spare cash to pay for life insurance for a director or players other than by accident, which has already been mentioned.

Theft is an ever-present factor. Make a list of every valuable piece of equipment and insure it, then lock it up at night in a safe place. Lock vehicles during shootings.

Another very important item is negative insurance. If negative material can be proved defective, the manufacturer will replace the bad material and that is all. Rolls of films can be lost, strayed, or stolen. Laboratory technicians will sometimes spoil a roll. They are human, after all. The cost of the material is negligible; the production time is 24K gold. You can and should insure this loss, because it's one of the most important parts of your whole insurance package. The negative is the whole package.

When you finish the insurance chore, the start date of the picture is staring you in the face like the muzzle of a Colt 45. Calls are put out to all personnel and players; check on the whereabouts of everybody concerned. This is the day before the battle. All rental equipment has arrived, and must be checked out against the order list, because tomorrow will be too late. Item by item, all the production work done up until this time is checked. This is especially true of location filming. Many times you don't have a second chance. Test-run the cameras, recharge the battery belts, and have everything ready to go. You will be surprised that, even with this careful checking, you will find tomorrow that you need this or that small item. That's what petty cash is for.

Long before this time, some will ask, "What about screen tests?" Screen tests simply aren't made on this type of picture. There is no time, money, or personnel available for them, and they very seldom tell you anything you don't already know.

On one picture the star was simply missing until a few hours before the production started. Phone calls, messenger, and carrier pigeon failed to reach her anywhere, and worry and gloom spread thick over the production personnel until she arrived.

Of course, shooting may start without a given person and some arrangements have to be made. Consult your breakdown and shooting schedule. Find out where you can logically start without upsetting the entire production. Look for scenes at the same location with the actors you have available, shuffle your schedule to place the scenes of the missing actor at the end of the sequence, giving the producer time to locate the person or, if it is at the beginning of the picture, to contract another player. These changes can usually be carried out with only a brief delay if your breakdown is very well-prepared.

Do not think the last minute rush is unique to your production; it happens to one and all. Just be sure it's an orderly rush using lists, material, and common sense.

SUMMARY

- For most low-budget films, filming in a sound stage can be prohibitively expensive. Your best bet is often to find a real estate agent specializing in furnished rentals. Get property releases.
- Know your facts before you start shooting. For example, some parks charge filming fees. Are you in one?
- Be sure to get the proper insurance coverage to cover such things as worker's compensation, property loss, negative loss/damages and personal injury.
- Place the calls to all personnel and players. Know where everybody is, where all equipment is, and what you're going to shoot first.

8

The First Day of Shooting

This chapter is oriented towards the director because now that all the production forces have placed the orchestra in front and the audience behind, the director alone must take the podium. Everything that could be done has been done and all the changes that will be required in sound and light programmed, so now it is up to the director to lead, conduct, and direct. Just like all the others, this is the director's first time, too: a new picture, a new crew, a new cast, a new story. The sensation, the fear, the joy, the challenge, are all there when the first slate goes "clack." It's thrilling — inexpressible — another picture has started!

As you will remember from Chapter 2, the director was called in to aid in the elaboration of the breakdown, after taking into consideration the first day's work, reading the script, studying the breakdown, and putting things in order for the shooting schedule. This first day's schedule should consist of 25 or 30 setups of a relatively simple nature, all related to the first sequence to be shot. This material should be neither highly dramatic nor technically very complicated. Even the most hyperactive action pictures contain these easy shots, which should be done in the beginning at a rapid pace.

The director has to start promptly shooting the first takes, and in 20 minutes they should be in the can. Without further ado the director has to walk over to the next setup and say, "Let's take — 124B; that is right here." At the end of the day at least 25 setups of two takes each will be done and the producer will be happy. The crew and cast will be tired but satisfied that the picture has started well. Some will say the director is a driver, and rightfully so. But there are many good reasons for starting a picture with this pace. The director has to establish from the beginning the rhythm or work pattern, and the pace of the shooting. He should not be domineering, but simply do it. If questioned about this, which can happen, no lengthy explanations are necessary but a reminder that it is a low-budget picture.

Recently a low-budget action film was started with highly dangerous and difficult car crashes. At the end of the week the producer replaced the director with another because he wasn't turning out enough work. Hard feelings all around thus prejudiced the picture. Whoever scheduled the work simply started at the wrong take. Had the less dramatic scenes been scheduled first, after a couple of days of rapid team work, everybody would have been in condition to start the more difficult part of the task. As this example illustrates, highly dramatic scenes must not be scheduled until the cast has had the opportunity to get used to each other and to fully understand the tone of the picture and what is really going on. Working in this fashion is not a luxury. It gets a lot of important shots done that are just as vital to the picture as the high drama or dangerous stunts.

On practically every picture, the team will face a dramatic and quite sensitive subject, seldom if ever discussed. At some point, usually close to the beginning, there will be a challenge to the authority of the director. It can come with rapidity and from any direction — crew, cast, or production — and it must be faced calmly but head-on, with no evasiveness whatsoever. Let me tell you how it came to one director.

The picture has begun at a good pace. The director had hacked through about 20 setups and was on the second day's work when he was called off the set for a brief moment. When he returned, the cameraman and the star were setting up the shot, placing the camera, ordering the lights, etc. This director, usually a furious type, got hold of himself and went off to another room of the set.

In a few minutes the cameraman called him. "We are ready to shoot, sir." The director strode onto the set, looked all around and said, "OK. The camera goes over here. 'You,' he said to the star, 'come here.' Now let's change those lights." The cameraman protested, "But sir, it's all set up." "I'm sorry," said the director. "The camera goes over here." The actor commented loudly, "But I thought . . ." "Yes," said the director, "the camera goes over here with a 50mm lens because I want this image size to jibe with the following shot." The actor stammered, "Well, I don't know about lenses." The director said, "Well, I do," and turned away to finish the setup. The cameraman didn't say a word.

The challenge may take other shapes and forms, but it's the test of their new director. A director who is the boss will prevail; if not, the picture will be a mess. There can be only one captain of the ship.

Sometimes the problem will be a petulant player who doesn't like the angle. Another day the executive producer will say, "Should we not do it this way?" The director thinks, "We? Who in the hell has to sign his name to this picture, anyway?" but will answer with diplomacy, "Don't you worry, it's all worked out," and turn away to something else. The most important thing for a director to remember is not to get bogged down in senseless, time-consuming discussions. No one dares question a symphonic director's reading of the score, and no one has the right to question the film director's reading of the script. However the objection is presented, it must be met with diplomacy. The best way is to keep right on working. These challenges seem infantile, and usually are, but anyone who has commanded people will recognize one for what it is and take it in stride.

At this time we should put in a few words about "committee" pictures. This type of effort may be useful in putting together and finishing a picture. Let us tell you of one such effort.

Several friends bonded together to produce a feature comedy and everybody helped everybody else. Six months later they finally finished shooting, after several cameramen and three directors had participated, the last one getting the job and the screen credit for "supervising a sequence." About a year later, the postproduction was finally finished. As a report went on, every scene was planned, discussed, and shot at leisure. Crowds of friends and acquaintances were pulled in for "mob" scenes. By any reasonable standard the photography should have been finished in 4 or 5 weeks.

The picture was presented at a private viewing and it was an average home movie. It was never distributed or exhibited commercially — most "committee" movies are never exhibited because most of them are deadly boring. As everyone has his or her own point of view and wants it represented in this type of production, the picture becomes flat because there is no single, coherent interpretation. Any effort to please a number of bosses is never as good as an individual effort, even if it is flawed. In one way or another, bureaucracy in any form produces only mediocrity.

Slates are the key to printing the chosen takes and editing your picture. They should be as large and clear as possible. The slate should be read off aloud before each take: "One-fifty-three, take one."

It serves as signal to all concerned. If, for any reason whatsoever, it's impossible to take the slate before the shot, hold it upside-down and take it at the end of the shot. Sometimes when it's necessary to stop and start the camera several times without any opportunity to slate each take, put the slate before and at the end, upside down as indicated, and so note it in the camera report that goes to the lab, and have them print all the takes between. There is no practical way of separating takes of this sort.

At the end of the take, the script supervisor has to make adequate notes. The director must check off the number in the script and put an X in the appropriate column of the breakdown. The time and footage should be noted with the magazine number and, of course, the reports should be dated. Later the editor may need all this information to locate a vital piece of film, usually an outtake that has to be printed because the first choice for some reason was unsatisfactory. The director will need the information later that evening to check the day's work and lay out all the work for the next day. The screen time is very necessary because it tells the director how much time has been shot relative to the total calculated running time of the picture.

A case comes to mind of a low-budget director who was shooting fast and furious to finish a picture, which he did in record time, but he was careless about the elapsed screen time he had shot. When the editor went to work on the usable takes, he found they has a 60-minute picture. More sequences had to be written and all the participants called back to get another 20-some minutes to complete the picture.

In the old days of the industry, complete pictures were made in two reels, but to be inadvertently a two-reeler short is quite a different thing. Of course, the above-cited example shows that this picture wasn't planned correctly from the first stages of scripting and breakdown. It can and does happen that when the filmed material is edited there are missing parts and mismatches.

All of this timing, check-offs, reports, and apparently redundant paperwork helps the lab, the editor, and especially the director. One of the director's most difficult tasks is to stay on schedule, and you will encounter this from time to time as many different aspects of work influence its rate of progress. The director is the only person responsible for keeping on schedule. If the cameraman is slow, it's the director's job to hurry him up; if the actors are having trouble, it's again the director's obligation to alleviate the problems by explaining the scene more clearly. Staying on schedule means taking advantage of every minute. If you divide all the minutes of a production by the cost to end of principal photography, you would be surprised at the amount involved.

Lay persons reading stories in magazines, newspapers, and books think of moviemaking as an exotic affair, where great dramatic decisions are made on the spur of the moment by sheer genius of producer, director, or star. In reality, it is hard, sweaty work, done bit by bit, with the intense concentration of all concerned. For example, few people stop to think, as they watch the ballet *Swan Lake*, that the dancers are working to the limits of physical endurance to create an illusion that what they are doing is effortless, and that is as it should be. Work behind the screen is similar. Work, sweat, and inspiration produce an illusion to gratify the senses.

Filmmaking is hard work, boring paperwork, and stress. Then why do low-budget producers do it? The satisfaction of viewing the works of their own creation, and seeing the final product on the screen, the combined, coordinated result of so many hands, hearts, and minds, is unexcelled.

SUMMARY

- The first day of shooting should consist of 25 to 30 setups of a relatively simple nature. A smooth and rapid start satisfies the producer, the director, the cast, and the crew.
- The director should handle challenges to authority quickly and diplomatically.
- Slates are the key to printing the chosen takes and editing the picture. Make sure the writing is large and the slate is read off aloud before each take.
- Be meticulous about the amount of screen time elapsed. The editing room is not the proper place to discover you're 40 minutes short of a full-length feature film.
- It is solely the director's job to keep the filming on schedule. Keep it on schedule. Time is money, and low-budget films don't have a lot of either to waste.

9

Directing the Actors

Aside from the few well-known books by theoreticians of the cinema whose feet float above the earth of us groundlings, very little has been written on the subject of directing motion pictures. This is quite easy to understand, as directors deal principally in abstraction. The very act of trying to describe it requires the translation to another medium. It is like trying to describe in two dimensions, that which takes place in three. This, of course, does not mean that the mechanics of the craft should not, nor cannot be explained. It can be done, but the sensitivity of the moment is lost in translation.

Direction begins with the reading of the script. As most directors are visually oriented, the words on paper conjure up in the mind the final image. The director can hear the characters speaking the words and moving through the scene. The basic interpretation of the story begins at this moment. This mental imagery will later on be translated into real imagery before the cameras. The director must be so familiar with the script that the mood, feeling, and expressions of the characters are almost palpable. They are real to him even before the parts are cast.

Of course there are adjustments to be made when the physical counterpart of the character appears before you in the flesh. Choosing an actor is, in great part, determined by this preconceived image. Of course, if the face and figure of an actor is already known, the director mentally places him or her in the scene. This is a conscious form of projection, rather than purely responsive instinct, as when reading a novel.

Most plays for the theater and screenplays for motion pictures represent only the bare skeleton which the director places the raiments upon and surrounds with atmosphere. This kind of literary effort must be read with determined creative imagination.

The transaction of reality begins with the readings for the principal characters, and is carried out either to choose the players for the parts or to familiarize them with the roles. Sometimes it's difficult to get all the players together at one time, but it is wise to make an effort to do so. This saves repeating the same explanations time and again. Each player should then read the entire script. Lack of time to do so hinders the comprehension of the psychological makeup of each character. Even the most trite plot and the most ordinary character can be given additional dimensions by a well thought-out interpretation. "There are no small parts," someone once said.

In these sessions, the director pinpoints the salient characteristics of the roles and instructs the actors as to their subtleties. The few hours needed for this work are well spent, for the actors can understand the basic desires of the director and now know in which direction they are headed. If left to themselves, their readings of the parts may be in complete variance with the concept desired by the director. After this orientation, the parts will be fine-tuned in rehearsal on the set.

Professional actors require direction just as any other type of actor. Take, for example, an actor who has just finished a long-running, successful stage play. The ingrained characteristics of the former role tend to carry over into the present one. Mannerisms not appropriate to the new role are involuntarily present. The characteristics of the new personality to be presented must be stamped on this ductile, moldable material of a sensitive actor.

Many B movies were badly acted, not because the principals were bad actors, but simply because they were not brought consciously into the depth of the role. This precept may seem to run counter to the simple fact of making a low-budget picture. It is not, because, for those few hours that you spend explaining to the players the basics that you desire, you may save time on the set. Our efforts should be directed to making the best feature we can make within our limitations, not simply to make a picture. Professional performers are quick to sense the difference, and will respond to the challenge.

Working with nonprofessional actors presents a completely different panorama. Nonactors simply do not know what is required of them. They are expected to know what they have never had the opportunity to learn. The director must spend more time in rehearsal to get a good performance.

All directing takes the form of suggestion — a very subtle form of hypnosis. Every word, every action, and every intonation should be a suggestion to the performer as to how to portray the role. There is a great difference, however, between suggestion and acting by the director. These suggestions should *never* be acted out. This is what the director is trying to elicit from the player. And this is the problem with using nonactors. Nonprofessionals simply do not know where to start. If left to themselves, they simply don't move.

First the physical movements should be explained to the players, because acting a movement physically leads to psychological feeling of the movement. Then with gentle, confident suggestion the meaning of the scene can be impressed upon them. Scolding or shouting in desperation only leads to nervousness and confusion of the player. There is precious little time to teach the basics of screen acting to a nonprofessional on a low-budget feature, but sometimes it has to be done. Patience and confidence are hard to come by in the highly tense atmosphere of this type of production, but you can maintain them if you try.

Much has been said about the "method" actors that crack their fingers and minds, preparing themselves mentally and physically for the maximum depth of interpretation of the upcoming shot. There is no time for transcendental meditation on the set of a low-budget movie. Good professional playing and understandable language are necessary for this type of picture. Many players have been taught these methods by their favorite teacher. Such techniques rarely have any relation to real acting, either on stage or in front of the cameras. In essence, there is no infallible method to great acting. Persons are either born with the ability or patiently learn it, most by study, watching, observing, and doing, rather than by the use of some mystical approach.

To paraphrase a common dilemma, "How can I get experience in directing if I can't get a job directing?" Without a doubt, the best place to learn acting and directing is in the theater, any theater. For thousands of years the theater has provided an arena for playwrights, actors, and directors. Today there are more of them than ever. Every town has a little theater. Colleges and universities have them. Larger cities have many. In those theaters you will learn what to do and what not to do and you'll enjoy first-hand the audience response and their applause of appreciation, as well as their rejection of the mediocre. This great crucible is available to everyone.

Then read and read. Read plays from the Greeks, Romans, French, Germans, Italians, Spanish, English, and Americans. Study everything you can get your hands on. Then go to the theater and to movie houses. See from the front of the house how others did it. Rent video cassettes. Run and rerun them. Stay away from the vast epic productions where special effects make the movie, because, on a low-budget picture, these are difficult to make. Rent movies about people and study each character closely. This knowledge, tempered by your own experience behind the curtain or screen, will build a fund of practical experience. The mine of material is inexhaustible.

As you read the plays of great playwrights, think how you, as a director, would block out the action. Think about what you say to the actors: what would your instructions be? How would you like to see this up on the big screen? When would you use camera movement, player movement, the long shot, the medium shot, the close-up? How would you see it? How would you do it? Ask yourself all the questions; then, in your mind's eye, come up with all the answers. A good director must have the ability to conjure up any scene just as he or she would film it on the set.

Not everyone is so fortunate as to begin in the theater and progress at an early age to professional motion pictures. If you are lucky enough to work in motion pictures, in whatever humble capacity, keep your eyes and ears open and observe intently the players, camera operators, and directors. They are experienced and are doing what you want to do. Observe, but do not criticize, because you may not know the rest of the story from your position and remember, very few ever started at the top. The great Hitchcock began in editing, passing to camera operator and then to directing.

A low-budget picture director has to know as much as possible about everything concerning production: how to film a scene as a cameraman, in case of an emergency; how to edit and cut, in fact, not just in theory; how to act, even if those days are long gone. There is no road to instant fame as a director; if you don't know enough of what you are doing you can fall into the abyss.

The word director means leader, and to occupy this post you should know every note of the score. Though proficient in one instrument, you must know the capabilities of all the others. Also you must possess the strength and sensibility of leadership and be willing to pay the price, which is hard work, long hours, and sometimes even doing the work of others. The rewards, however, are so great, the price paid fades into oblivion.

There is more to being a director than simply saying "I am a director." You must demonstrate your capacity to direct and be prepared as well as possible to do the job. You must inspire confidence by knowing what you are doing, and doing it well.

Of course, there is a "how" to everything, although the physical how has to be backed up with the mental how. Let us say, for example, that two men are seated at a table with a good looking woman between them. One is an older, experienced man; the other is young and ambitious. The woman is apparently not aware of the implicit rivalry. As the older man talks of his exploits, the younger one, irritated, stands up, goes to the background and lights a cigarette, waiting for an opportunity to express his opposition. He does so, and the scene terminates where the other two stand up, the woman conciliatory and the older man lightly ignoring the younger one's statement. This simple sequence would be divided into several takes showing the players in different setups. It might be done in close-ups of the principals to demonstrate their actions and reactions, finally to return to the original setup that established the scene.

This means the director must decide how to display the action to the audience (see Figure 11). There may be several ways to do it, but the director must transfer the concept to the physical counterpart.

▶ ***Figure 11*** *Just prior to the take, the director instructs the actors on the set regarding dialogue and movement in a scene with underlying conflict.*

If necessary, use chalk marks for start and stop points. They are invaluable in providing focus pulling and to obtain consistence in acting. A good actor will hit his marks every time.

A word at this moment on scene "stealing" would not be amiss. Players often try to intentionally steal scenes from the other involved. This must be avoided at all costs. It must be put down firmly. If not, the director will lose control over the most vital part of the picture: the complete statement that the director wishes to make, rather than one the actor would like to make. Some players think they are getting away with stealing a scene, but the director and the editor can always cut anywhere they choose and, in effect, nullify the effort of the player.

A direct form of dealing with this problem is for the director to simply take aside the player over whom the other wants to prevail and tell him to make some specific movement that will distract the audience completely from the cheater. The director always has to have the last word.

As the actors have read their script and know the lines, the director blocks out the action of our little ménàge à trois just as in the theater. We will assume that he or she has already given the indications to the cameraman. The indications to the older man are that his mood should be self gratifying, expansive, and happy; the presence of a beautiful woman at his table listening to him tell his exploits makes him content. The woman is hearing, for the first time, tales by a real adventurer and is rapt. The director's instructions to her are to be attentive to the words of the older man and to participate in his mood. The instructions to his rival are to convey a sense of suppressed frustration and rivalry and, when he can take no more, to get up and retire to the background, placing himself visually between them as seen by the cameras, lighting a cigarette, and impatiently awaiting his opportunity to compete with the other man.

After the blocking and tentative reading of the lines, the director makes small changes in emphasis and intonation. Remember, lines are not just read by rote, but should be given just enough expression to be valid screen acting. (A little later we will go into the differences between screen acting and play acting.) After these are run through, the director turns to the cameraman and says "Ready? Camera." The cameraman responds "Yes," and the camera starts to roll. The person with the slate steps in and announces in a loud voice, "152, take one," clacking the two parts of the slate together. "Action," says the director. After the action is concluded, the director waits a brief moment to catch the reactions of the players and calls "Cut."

By the way, the proper position of the director is as near as possible to the left side of the camera. One who knows lenses will know how area much is covered. The actors look at the director, who steps forward and says to the older man, "Wait just a bit before turning to her for your laughing answer." "Now," stepping over to the young man, "be sure and hold still while he answers." To the girl: "Tilt your head just a little to the left when he comes on strong." "All right, ready? Camera." It is slated again. "152, take two."

After the scene is finished, the director marks the script and breakdown and calls to the script supervisor, "Print two, please." Now let's go to the close-up, 152-B. The director talks earnestly to the young man about the close-up as the camera moves in, and the gaffer starts lighting the close-up. Thus it is, each piece, each bit, each word or sentence is worked on to get the maximum interpretation at the best pace possible, from 9 A.M. to 1 P.M. and 2 to 6 (or 8) P.M. with two coffee breaks if you're lucky.

The relationship between the director and the actors takes on a quality known in psychiatry as dependence. The director is a parental figure whose sternness disapproves bad workmanship and whose approval of a good scene delights. Like any good parent, the director should play no favorites, but treat everyone the same. Actors are very sensitive, sometimes insecure persons who, underneath their exterior bravado and apparent sophistication, require supportive direction. They must be treated as children, but not childishly. This simply means that, in front of the cameras, like musicians in an orchestra, actors are subservient to the will of the director. Like any parent, a good director will not mistreat his children, but sometimes they have to be persuaded, cajoled, and even forced to do the parent's will. If no evil intent is pursued and such manipulation is for the good of the picture, it must be done.

The consideration of dramatic values is always a difficult matter. Today there are basically three different styles of acting, and the differences between them must be thoroughly comprehended.

First there is TV acting. In spite of the enormous quantity of work done every week, the general quality is not bad, considering that an entire episode must be shot in five working days. Most actors simply read their lines and, if they are intelligible and make some sense, that's it. This leads to a certain flatness of delivery. If you have the opportunity to screen several of these shows in succession, the result becomes perfectly clear. The lead depends upon his or her personality, and the surrounding cast members are usually stereotypes. Don't be surprised if some low-budget pictures sound the same as others. This flatness must be corrected by allowing more expression for the actors.

Next we shall consider stage acting. Here the audience stretches from the first to the last row. Even in the smallest of theaters some projection of voice, some exaggeration is necessary, otherwise the performance will be flat. As a director you will find your instruction running thusly to seasoned stage actors. Because they are used to the stage, they have the tendency to overdo it. The advice is "think the part and it will come to you naturally. Remember, the big screen exaggerates everything." The actors then become introspective and turn out a terrific performance because they are holding inside them the necessary feelings; on the big screen it comes through as great acting. The trick is to build it up, then hold it back.

Screen acting is a very special type of work. It's not as simple as Pudovkin's intercutting of scenes. Juxtaposition does have something to do with the overall picture, but nothing with acting. Think for a moment that the size of the TV screen will never compare with the movie screen and will never completely overpower the movies because of this fact. The sheer size doesn't even invite comparison. In the same way, the movies never finished with live theater, owing to the physical presence of the actors. Each medium has its own sphere and will always have its own following.

Motion picture acting has a relation to the screen size. The thoughts behind the expression are as important as the expression itself. That is why hypnotic-like suggestion is so important in screen acting. That is also why a complete understanding of the character being portrayed is so necessary, even though the picture is shot out of sequence. The screen director alone should decide exactly how much emotion is required at a given tine. While some actors may object to the director's interpretation, their objections must be overlooked and their emotions controlled and played upon to bring out the subtleties that the huge screen requires.

Timing is one of the most required aspects of screen acting and it behooves the director to understand it. Screen timing is different from that of the stage and even more so from that of TV. For the big screen, physical movements are spread across many feet. Dialogue comes from heads 12 feet tall. To rush this causes strange effects. Movement, while not being slow or lethargic, should be more thought out and precise. The timing of speech patterns is also different for the different media. Diction should be clear and not overdone. The opposite, mumbling, in spite of the style of a celebrated actor no one can understand, is definitely to be avoided unless called for in the script.

Just how to do all this movement and enunciation without appearing stilted is the trick. It is done by thinking about overdoing it, then holding it back, suppressing the urge to rush across the stage. Those dramatic scenes where all the stops are pulled out have to be very well orchestrated to avoid being completely overblown and overdone. Motion picture acting must be close to life itself.

The director must constantly be aware of timing to avoid a choppy picture, as correct timing gives the whole a fluidity much to be desired. Do not confuse this timing in direction and acting with timing influenced by editing. That is another matter entirely.

As you will have noticed, the screen director in the double function of blocking out and directing the action and the actors, also controls the amount of production work done in a day. Refer to the sample breakdown in figure 4, page 22. Nine different setups are specified, a little less than a half a day's work, but with the heavy love scene it may run longer. To compensate for this time, the director devises a shooting schedule for the next day, calling for 25 setups in all. The night before the shooting is the best time to lay out the order in which the work should proceed. You may start with #44, #46, #72, followed by #45 and #51, and so forth, taking into consideration the positioning of the camera, lights, players, etc., to do the work with a minimum of change in setups. This doesn't mean, however, that you will use exactly the same position or lenses, because it is monotonous to use exactly the same camera positions for all these setups.

This preprogramming is quite important to save time on each individual setup, possibly allowing for two or three more shots at the end of the day. This prepared work will leave the director's mind free from these matters while on the set, where more effective work with the actors can be accomplished. By looking at the eighths of pages, workload can be timed out judiciously, noting this one is easy, but that #74 is going to consist of many takes. Number 74 is a love scene slated at the beginning and end on a closed set. It will take more time and sensitivity to accomplish well.

The physical union between two beings that creates the life of a third and gives them pleasure in doing so is sublime. The fact that its practice can be considered bad or dirty is absurd. But, as in all things, there is a question of taste in presentation. The old days of an embrace followed by crashing waves are gone forever — or are they?

Taste, style, flair, whatever it may be called, is an inherent part of the individual. Let us set up a hypothetical sequence. We are in the studio. The lights are there, the camera is there, and there is a very beautiful woman, a physical ideal — completely nude. In comes director A, who sets up the lights, poses the woman, and creates pornography. He takes his film and leaves. In comes director B. The same lights, the same camera, the same woman, completely nude. This director sets up the lights, poses the woman, runs the camera, and creates a work of art. The difference consists not in what was done, but in how it was done. Any sexual act can be presented either with taste, affection, and appreciation, or with extreme vulgarity.

In spite of the fact that one X-rated porno movie has made its producers many millions of dollars in the last few years, this type of presentation is not appropriaate for general distribution in most of the world. This brings us to the question of ratings, mentioned earlier.

There is a very generalized system specifically set up by the Motion Picture Association of America, not to protect the public, as commonly thought, but to protect the producers, distributors and exhibitors of motion pictures from the thoroughly misguided efforts of some special interest groups to impose their own brand of ideas through censorship on the screen. In spite of protests from small groups in the motion picture industry, the present rating system of G, PG-13, R, and X is a fairly satisfactory compromise which covers the range from family programs to porno fare in an equitable manner. Low-budget film producers under normal conditions should try for an R, although there are exceptions to any rule, as we will see later. A new rating of N-17 has replaced the X rating, but the X will be long remembered..

The secret of successfully portraying sex is that it should be part and parcel of the story, and not pulled in from off-stage by the hair. After this is taken care of in the scripting stage, the director must have sufficient good taste to shoot the sequences in such a way so as not to offend the mores of his or her chosen segment of the public and still have the film fall into the proper rating category. This is no guarantee, however, that the picture will receive the same censorship treatment outside the United States. Rest assured that you cannot satisfy everyone, but do try to get the appropriate United States R-rating. How far can you go? You can come awfully close to an X-rating if, and this is a big if, it is very well done. The best way to find out is to look at other producers' films and ratings at the time of production. This work must be recent. The Motion Picture Association's thinking as to what goes and was doesn't has changed slightly with the passage of time. Try to be up-to-date and don't get locked into a time frame from the past. Pictures do date, you know!

Looking at the practical aspects of the shooting, it will be seen that they are quite simple. Use a medium shot for full-length scenes as an establishing shot, then move in for close-ups of important pieces of action. A hand-held camera is useful at these times. Hopefully your cameraman won't develop the shakes. You are always safe with full-length rear views and waist-up close-ups or over-the-back and shoulder shots. Give the editor plenty to work with in angles and viewpoints. A few minutes spent taking them may make or break your picture, as these are dramatic highlights. Don't forget to include a couple of big facial close-ups of ecstasy, even though the camera pulls back later. They are vital to success. Let's hope you enjoy directing sex scenes as much as you enjoy sex. They are the culmination of attraction between two opposite forces of nature representing love, passion, and the inevitable. As a director, make the most of the opportunity presented to bring your public right up to the fire of life.

Your stars will feel much better toward you as a director and give better performances if you film all your sex sequences on a closed set. A closed set means just that. Everyone out except, of course, the director and the camera operator. With a feeling of privacy the players can come up with better performances and follow suggestions that couldn't be made openly without embarrassing them. Scenes shot this way take on a special feeling of intimacy that is almost impossible to obtain in other circumstances. When you come off a closed set, the curious will come up and ask, "Did they really do it?" Never tell them anything. Imagination is much more powerful than reality. We live in a world of illusion.

Human beings are violent animals; when considered from a purely intellectual point of view, the cruelty of people to their fellows is simply incredible. But most members of the human race are not intellectual and don't even stop to think about it. We are horrified at the sight of a man killing a bull in the bullring and forbid it to be exhibited publicly, but we permit or simply watch thousands kill other thousands in wars based on primitive philosophies, or see people starving to death in one part of the world while food rots in warehouses in other parts. Now this does not mean that one prevalent dogma is better than another. It means simply that at the present the problem has not been solved. Since the problem of violence has not been solved on an international scale, do not expect filmmakers to solve it on a modest, low-budget picture scale.

The presentation of simulated violence on the screen, and it is always simulated, is quite different from real-life violence. Real blood all over the place is not the same as synthetic blood sprayed around. A man with his guts blown out in battle is not the same as the simulated scene. Real violence is more brutal, nauseating, and disgraceful than anything seen on the screen. Violence did not begin on the screen; our pictures mirror our times.

Before loosely criticizing filmmakers, we should, one and all, criticize ourselves. As long as we must live with this dichotomy in people, if we are to present a true reflection of them, we must at some time present it on the stage or screen. Euripides faced the same problem. Violence for its own sake is not enough. In any good picture that is well-written, the violence should be an integral part of the plot and not gratuitous or misused rather than used.

The director should study the violent sequences of the screenplay and make notes on the most effective way to present them to the viewer. The traditional ballet of the barroom fight in the classic Western picture has no relation whatsoever to a real barroom brawl, which probably would not be spectacular in the cinema. Our violence in pictures is almost always as stylized as the Japanese Kabuki theater, but it is not perceived that way because our art form is familiar to us, and accepted as a true depiction of life.

All violent fight scenes should be well studied and choreographed with the actors beforehand. Usually the night before shooting the sequences, all the major moves should be worked out on paper with a copy to each combatant. The fight must be practiced several times so they will get the feel of the sequence. If the cameraman can be present at this session, so much the better.

When the time comes to do the sequence, be sure to have two cameras on hand for the master shot. Be sure to load the same type of film stock in both cameras. Ideally the rolls should have the same batch number. This will avoid jumps in color values when editing. Use a semi-telephoto 100mm on one camera on a tripod, and a 50mm on a hand-held camera to move quickly from one shot to the next. Have your makeup person ready with all the material to apply between takes. A spray bottle of water for faking perspiration is a must. Be sure to get some frontal attack shots of each actor as they charge each other, a long shot, and close-ups of dramatic faces and torsos in violent action. In this type of sequence the camera must get right into the action. Your orders to the actors should be to make it as real as possible without damaging each other too much. A word to the wise: Program your violence at or near the end of the picture and be sure you have scenes with other players to film afterward. Violent scenes are very tiring to your actors, they won't be at their best after such work, and there may be one or two quite apparent bruises, battered spots, and scratches, even in the best of circumstances. Be sure you have duplicate costume changes as the ones used will be ruined, and keep the first-aid kit handy.

Problems will arise for the director on every picture. In some cases the director must literally brainwash the player with suggestions to come up with the special effect required. Sometimes actors in the same shot must be given different instructions separate from each other. This is done to provoke spontaneous reactions from them. It's a form of cheating and is valid for good direction.

Let us get back to the technical business of directing. There is a simple, comprehensive, all-inclusive dictum: What shows up on the screen should be exactly what the director wants the public to see, chosen by the director alone. He or she must know which lens covers the area to be seen, the speed and movement of the action, what different lighting setups do to the picture, and if the take will cut with adjacent shots, and whether he or she is crossing the line or not, or calling for a false reverse. Then the director calls the shots, directs the actors and supervises the editing, and in an emergency, should be able to carry out and perform the duties of the cameraman.

In spite of what a famous director said, you must look through the viewfinder for every shot, and also at the beginning and end of every camera movement. There is nothing denigrating in this. You are not admitting that you don't know your lens coverage; you are simply certifying what you have called for: the framing of the shot.

SUMMARY

- Directing begins with reading the script. The director must know the script backwards, frontwards, and sideways before the casting even begins.

- Try to pull the entire cast together for the initial readings of the script. It saves time in repeating the same instructions and familiarizes the full cast with the entire script.
- All directing takes the form of suggestion — a very subtle form of hypnosis. You should *suggest* how to play a part, but *never* act it out yourself.
- As a director, read everything you can. Practice blocking movements and directing actors and actresses. Rent movies. Study the blocking and the acting. What choices were made that you agree with? How would you have done it differently?
- Know what every member of your crew does.
- Make up in your mind what each scene will look like. Know the area each lens covers and what each light and effect will look like on film. Only then can you properly block out each scene and direct the action.
- Don't be afraid to use chalk marks on the floor when filming. Good actors can always hit their marks.
- There are basically three different styles of acting: stage, television, and screen. Stage acting is exagerated and flamboyant. Television performance is somewhat flat. Screen performance is somewhere between the two, and most closely mimics the actions of everyday life.
- Timing for a motion picture is crucial and must not appear rushed.
- Love scenes should be tastefully portrayed. An X-rating is usually not in the best interest of the film. These scenes should be done on a closed set with as much sensitivity to the players as possible.
- All fight scenes must be carefully choreographed and studied. Practice the scene and be sure to have two cameras on hand at the time of the sequence.

10

Directing the Cameraman

Chapters 9 and 10 are the two sides of the coin. One deals with the basic elements of directing the theatrical or acting — in other words, what is to be filmed; the other deals with how to register this very complex matter on the negative material in exactly the way you want the viewer to see it on the screen. To do that you must take into consideration the position of the camera, the type of lens to be used, the angles, the kinds of shots you are going to make (either long, medium, or close-up), and the camera movements involved.

Camera movements nowadays are easy to come by. Right now we are going to shock the purists out of their minds. If you look back to the equipment list on page 52, you'll see a 25-250 zoom lens or its equivalent included. The primary reason is that it's a multifocal lens that you can use consistently to change the framing on the same line by a simple twist. No one objects to that. You can also use it for dolly or tracking shots. It takes hours of setups using a regular dolly; the zoom lens can save you from all those lopping hand-held or walking camera shots, and those bumpy wheelchair dollies that look so awful. Some will say the perspective is not the same; of course it isn't. But 99.99 percent of the viewing public does not know the difference between a well-made zoom shot, where the use of the zoom is not grossly apparent, and a dolly shot.

How many people out of the millions who may see your picture will care how it was done? Fie on the purists; we are in the business of making a good, low-budget feature as fast as we can with a minimum of sacrifice. This is a minor sacrifice at most.

Believe it or not, one of the most difficult shots to make is a simple pan. With a good fluid head the chances are excellent, but the camera operator must have a stable arm and the actor must hit his marks correctly. Mark the start and finish with a piece of chalk. Bad pans are the sure mark of an amateur. The camera should follow the action discreetly, and the movement should never, *ever* call attention to itself. Fortunately the early zoom madness is now a thing of the past.

Tilts fall into the same category and must be well done. A studio hydraulic model dolly for raising and lowering the camera is usually out of a low-budget production for many reasons, including rental cost, setup time, additional crew to operate it, and the means to transport it. Suitable substitutes have been used, however, such as wheelchairs and wagons.

Image size is always something that has to be watched carefully, as well as image location on the screen. This is especially applicable to reverses or over-the-shoulder shots. If there is a great disparity in location or image size, this calls attention to itself or, at least, produces a certain malaise in the viewers. They may not be able to put a finger on the technical reasons for their dislike or lack of satisfaction with a picture because the cause is subliminal or not apparent at first sight. Give the viewers every chance to like your picture and lose themselves completely in the story, rather than being distracted with technical errors outside their ken.

Before leaving this subject we should discuss the much-maligned *master shot*. Some directors boast of cutting in the camera, but that's like cutting off your nose to spite your face. Shoot the master shot then move in quickly to make all the niceties of close-ups. Why? If for any reason the separate shots refuse to cut together, without a master shot you have no choice but to correct the defect by using an inadvertent jump cut. Oh, we have all heard that jump cuts are modern and up-to-date, or that so-and-so "invented" them in Paris; but that is not the case. Jump cuts have been with us from the very beginning. Just because so-and-so used them to cover up his extremely fast or sloppy method of shooting is no reason you should use them. There are logical uses for jump cuts, but not as a daily fare. Abuse leads eventually to debased coinage. Give the editor and yourself an even chance by protection whenever possible. On low-budget films you'll have to supervise the cut anyway; producers very seldom do it.

Cheating, a low-down practice, is not even mentioned in most texts, but it has to be done practically every day on almost any production. The low-budget director must be as adept at this art as any con or short change artist is at his.

▶ ***Figure 12*** *For reverse shots this cabin cruiser was turned 180°, since this was the only angle with a clear view of the empty horizon.*

Here is a case in point (see Figure 12). In a certain picture the author, much to his regret, found that the isolated, exotic looking dock for the yacht was not available any more. Another one was located in front of a populated boat yard, flanked on each side by high-rise condominiums with an enormous concrete causeway and bridge to the mainland in the background. Less than 90 degrees of open water was left for unencumbered shooting, because the action was supposed to take place on a deserted island.

The director took one look and said it would do. Actually he didn't have much choice. It also bears mentioning the dock was a T-shaped affair with water on each side of the T, and the bottom of the T ending at a sea wall. Shots were taken of the players from the sea wall at distances of from 30 to 50 feet, using the zoom lens as a telephoto. Some shots were taken at a high angle cutting out backgrounds, some at a low angle against the sky. When reverses were needed, as they always are, the yacht was turned 180 degrees to use the only expanse of clear water as background. Naturally it was necessary to use plenty of fill light to obliterate the shadows that would have been a dead giveaway of the turning process.

As the players stepped off the yacht onto the dock, a cut was made to reverses filmed miles away in the dunes. As no extraneous elements ever entered a single shot, the illusion of a desert island was complete and absolute. More than 75 setups were made at this location alone, including the arrival of the yacht, all the interaction, plus the arrival of the U.S. Coast Guard! More than 67 setups inside the yacht were made at this same location, and at all times the feeling was one of lonely, vacant space and endless lagoon on one side and sterile sand on the other. This is directing at its best. Remember, the camera can see only what you want it to see.

Cheating on the set is another matter. Sometimes there is part of a lamp on one side distracting from the primary intent. The little part is not important, just move it off. Other times it's just turning the actor's head or stance a little to get a better effect and, as the objects within the scene are not visually related one to another, the change in angle is never noticed.

In another case that comes to mind, the door of the living room of a house did not lead outside. An angle was chosen that didn't afford a view to the outside and all the exits and entrances were made first on the set, then at a convenient door located elsewhere. In Western towns mentioned previously, the end of the street is almost invariably blocked off with period buildings to hide the fact that there's a glue factory just behind it.

In some cases cheating is carried as far as shooting scenes between actors who are not even on the set at the same time, and in some cases, even days apart. It may be necessary, but there's always a strange feeling that something is missing: the two actors together, the true establishing shot. Cheating with explosions, car crashes, or falling from a cliff are and have been the daily bread of low-budget shooting from the beginning of the industry. Never say a sequence is impossible until you consult your cheating heart.

Lighting the set is the province of the cameraman and the gaffer, with some assistance. The average motion picture set usually looks overlit, but really is not. The most laborious aspect of lighting is to maintain the correct number of foot candles of illumination on all the different shots and, at the same time, the proper lighting ratio or the correct modeling lights. Maintaining the correct quality of light avoids a larger number of changes in the printing process in the lab, as the timer tries to assume uniformity in a sequence. Jumps in lighting intensity are to be avoided unless expressly needed. A sequence leading up to a dark scene of storm with thunder and lightning, for instance, should be prepared in the previous related shots, then gradually get darker to avoid an abrupt change in atmosphere.

The ratio between key light and fills used to be three-to-one. That was when negative material didn't have the response it has today. Also when the old Technicolor 213 inhibition process was used, good dense blacks were easy to obtain. Nowadays much higher ratios can be used with Eastmancolor, new material so sensitive that it reaches into the darkest shadows. It's hard to get good, dense, dramatic black without resorting to high lighting ratios. The nice part of it is that the new material needs much less light in general and much less fill light. For this reason, 10,000 watts of quartz lighting, correctly distributed, will evenly illuminate a 20′ by 40′ set. What you gain on one side makes up for any loss on the other. If you are going to use some special new material, try it first and learn its peculiarities, advantages, and defects so you can be prepared for anything. This is easy to do if you own a 35 SLR camera. Take some rolls of this new material under different circumstances, send them off for development and see the results yourself. Of course you have no control of the lab corrections in this case, but it will give you a very good idea as to the qualities and limitations of the product as you have used it.

While on the subject of lighting, we should talk briefly about existing light. With new fast material you can get exposures with existing light, but most existing light is simply too flat for use with the new long-scale negative materials. It is usually completely lacking in dramatic quality. The best thing to do is to forget existing light and illuminate the set the way you want it, right from the beginning; besides, artificial lighting is more stable and dependable. Remember, too, that the laboratory can print up and down the scale to an amazing degree, going from day-for-night effects right up to blazing light, or the contrary, if you shot it on the correct part of the scale.

Color balance and filtering play an important part in good motion picture production. By the way, you can shoot 5247 without the 85 filter and use the entire available film speed, but you should keep these rolls separate and let the lab know about it so they can apply the correct filter in printing. If not, the lab people can get awfully peeved with you, and rightly so. Let them know any time you are going to deviate from normal procedures, and they will be very helpful about even giving advice where needed. If you are shooting inside and all your lights is controlled, you have practically no need for a color temperature meter, but if you have doors or filtered windows, you will definitely need one. Your cameraman will know when and where to use it if you make known your requisite in preliminary discussions.

Outdoors, incandescent fill lights must be filtered with Rosco blue to avoid some peculiar effects in the shadows under chins, noses, eyebrows, etc. Where existing light is under outdoor sheds, roofs, car tops, umbrellas, etc., this is especially true. Do your best to give the lab a properly exposed and filtered negative to work with. They then will give back to you the color corrections and densities you need. It all works together to produce a better print in the end. One of the most important uses of color in feature films is accomplished by placing color gels in front of the lights. An amazing variety of colors can be obtained, and it would be wise to get samples in the form of a swatch book from the suppliers.

It is possible to completely change the character of a set with the use of color. The changes can be so subtle as not to be consciously noticed, or so complete as to eliminate all but a single color.

The use of gels can enhance the quality of a picture without undue expense. This type of lighting is especially effective when used for night scenes or the interiors of bars and restaurants; needless to say, they are equally useful for theaters and ballrooms. There is only one small drawback: They need more light to obtain the same exposure, but this is a small price to pay for the many effects possible with their use.

Makeup must be considered under the heading of color (see Figure 13). It is a substantial part of the filming process and should be taken into account from the very start. Male makeup is almost never used — lipstick on men looks awful unless they come by it indirectly, such as in kissing women. Just be sure the men in your picture are clean-shaven, even if it has to be done twice a day, because blue beards make them look unkempt.

▶ ***Figure 13*** *An assistant applies droplets of water to look like sweat to the actor's face as the director and cameraman wait for him to finish. In this scene the man had just finished digging the grave and the woman is about to pray.*

The days when the leading lady wore wall-to-wall makeup and looked like a painted doll are in the distant past. Light, clear makeup with good, fine lip color and modest eyelash work is enough. Sometimes natural makeup color must be used on arms, neck, breasts, shoulders, and back to give an even coloring, but be careful that it is even and stay clear of fluorescent lip makeup. The best way to check makeup is to make tests beforehand with a 35mm SLR camera using 5247 negative material. Send it to be processed both as slides and prints.

To soften older features, you can apply several solutions simultaneously, such as makeup and soft focus. There are some new contrast filters that, used with discretion, can work wonders. A number one filter is usually enough, but don't fall into a trap. If you use the contrast control filter on one actor, you must use it on them all, otherwise you are going to have a very strange mix indeed. When you go to the medium shot, take the filter off. Another trick is to put a gauze filter on your quartz lights. They are hard to start with, but can easily be softened to a nice, soft photoflood quality with the judicious use of a gauze. Speaking of cheating, it has been bandied about that Canova, when commissioned to do a reclining nude of Paulina Borghese (Napoleon Bonaparte's sister), posed his beautiful young mistress for the body and sculpted Paulina's head on it. Whether it's true or not, it is neither the first time nor will it be the last that one body was substituted for another. It's done in pictures, too. In fact, in a recent big hit, the director supposedly used the body of another woman much younger than the star player in a gorgeous shower scene. Again, if it's true or not, only the director and cameraman know, and they aren't ready to talk yet.

Mismatches are the annoyance of the editor. The easy way to avoid them is to change completely the angle or the image size, and sometimes both. Cutaways are the life belts thrown over the side. Order those cutaways and preserve the editor's remaining hairs. Be sure to repeat the actions as closely as possible in the close-up so it can be cut properly. The worst mismatches are not in movement but in costume and objects. Be sure to get them in the right place. These things must be done right during shooting, rather than patching up the damage in the editing room.

The relationship between the cameraman and the director is critical in the sense that, while the cameraman is an extension of the director's eye and brain, the two must operate independently. A sense of harmony must exist between them in an effort to strive for the common goal. If the director is apt in choosing setups, angles, and movements, the cameraman will follow anywhere. If the director demonstrates indecision or generally is not sure of his or her craft, the cameraman, impatient at these proceedings, will take over command. In case of a strong-willed cameraman used to having his own way, friction can develop between them. These differences should be settled, preferably off the set, or at least out of the hearing of the crew and cast. A perceptive director can usually see these things coming on and deal with them in the same way he or she deals with the actors, utilizing tact, patience, and understanding of the other person's position and responsibility. However, if the director definitely knows what he or she wants, the real point of the discussion is to hear the other's suggestions, then make the judgment. To acquiesce to all the demands or suggestions is to lose control over the tone and direction of the picture.

The one person with complete knowledge of the overall structure of the picture, both artistically and technically, is the director. Although a suggestion may have great merit, it may not fit into the overall concept of the work, thereby creating a discordant note and, for this reason alone, must be rejected. This must be done with courtesy and tact, however hard these two qualities are to come by. The ideal situation between a director and cameraman is mutual trust and understanding.

SUMMARY

- Before you start, know the positions of the cameras, the types of lenses, the angles, and the kinds of shots you want.
- Don't neglect the master shot. It may save your scene.
- Many instances will call for you to cheat in creating a shot that looks like one thing but is really another. Be inventive; never say a shot is impossible until you're consulted your cheating heart.
- Lighting is crucial to the overall look of the film. Most film sets look overlit but they usually are not.
- Since it is possible to completely change the look of a set by the use of color, the color balance, gels, and filtering are also important. Keep careful records of each filter or gel used, both for the lab and for continuity.
- Natural looking makeup is generally best for both men and women. Gauze and special filters can be used to soften the effects of aging.
- It is vital that you maintain a good working relationship with the cameraman.

11

Viewing the Rushes

The rushes, the raw material from which a picture is assembled, are one or two days' work sent in a rush to the lab to be developed so the director can see how the takes are coming along and if any scene must be retaken. When the crew is still on the same set, retakes are fairly easy, but if they have moved to another location the retakes may have to be rescheduled on another day or at the end of the picture. Every effort should be made to shoot these retakes as soon as possible without disrupting the entire production.

Like all other product in elaboration, the work may or may not look very promising. It is like a diamond cutter, turning over and over an opaque piece of material, with no hint of the glitter or brilliance that will later come out of the gem.

Only the director and the editor should view the rushes. The omission of other persons is deliberate. On some productions, the producer, the players, cameraman, gopher, etc., all sitting there watching, are bent on making known their feelings. The actor grieves over his performance, the producer worries about the money being lost because the rushes look terrible, the cameraman is definitive: "This or that shot must be done over." Everyone would like to do over the entire day's shooting for one reason or another. Be authoritative. The director and the editor are the only ones to view the rushes, and if they insist, the producer, who, after all, is putting up the money.

The director and editor are the only persons who use this material to make up the picture, and they are the only ones capable of understanding what is to be done with the rushes. At the end of the week you should have a rough cut of some of the sequences which you must invite the producer to see. When the slates are cut off of a sequence and some scenes put together, the rushes will look quite different and the producer will be mollified. The production will save trouble and money by not having to reshoot sequences that can be salvaged in the case of minor flaws.

It is necessary for the director to have a portable tape recorder when the rushes are projected on the screen or viewed on the moviola. The tape recorder is usually handed back and forth between director and editor to record comments like "239 — let's cut right after he uncorks the bottle. Do you have a cutaway for this scene?" "No, but we can pick one up tomorrow." "Do you think that we ought to retake 357?" "No, it looks passable." "How about 358?" "If you can, let's retake it." "I agree." The give and take between editor and director goes on until the latter says, "OK, that is all for today."

Sometimes the director will have ordered prints of two takes of the same scene. The decision of which one to use must to be made in that moment. Sometimes it's necessary to tell the projectionist to rerun the lot to come up with decisions that are appropriate. The editor then takes the cassette and the can of film to the editing room.

Before cutting up the work print into "takes," the editor or assistant should run all the material through the moviola and check off all the numbers on the camera report as sent to the laboratory. The reason for this is simply that the printing ordered should coincide with the numbers on the report, because labs may sometimes overlook printing "253, take 3," causing the editor to go mad looking for something that's not there. If this number is missing, the editor calls the lab and asks them to print the number.

The editor must have a separate copy of the script and breakdown to get through the intricacies of putting the picture together in the proper order; the director, having written the numbers in his or her copy of the script, revises the shooting schedule.

The director must advise the editor when making changes via a copy of the script supervisor's notes. Occasionally a shot is simply missing, not having been taken at all. The editor must remind the director of this and insert a slug of white leader in place of the missing piece of work print.

Sometimes, after looking at a take which is not exactly satisfactory, the editor will consult the camera report and or the script supervisor's notes to see if another take would do better. If the notes indicate the take was complete and not rejected for major error, the editor may order a print of this outtake. Often even a defective take may be used if a cut can be made just before or after the error. It's amazing what a good editor can do with the material.

As the production nears its end, the director and the script supervisor should be reviewing the script, breakdown, and all the notes to see that everything that is supposed to be filmed has been done and accounted for. The director may indeed shoot some additional material to put under the main titles if a moving background is desired. Pickups and inserts must be shot before leaving the location. Everything possible should be done to insure that the principal photography is complete; later may be too late. If there are cutaways and some neutral filler material to permit the editor to work around abrupt dissonances, you as a director will have finished the most difficult part of your job.

All that remains to be done is to have someone check in all the rental equipment, piece by piece, and pack it well to return to the company from which you rented it, keeping a copy of the equipment list. Also note down all the burned-out bulbs, etc. If you do all this work with care, the rental company will consider you a good customer and be happy to serve you again. Store all the leftovers in the company warehouse and thank everyone, even the stinkers, for having cooperated so well in the production. Leave your location clean and orderly, even if you didn't find it that way when you arrived. Make a good reputation for yourself and your production company.

Now that everything has been taken care of and the last shot made, there is one more thing to do — the wrap-up party. As a director, make your appearance, take a drink, and a break away as soon as it's decently possible. They've had enough of you after a month of shooting. You can hear the results of the party secondhand the next day if anyone is capable of speech and close enough to tell you. Some of these parties are phenomenal, but you still have to finish the picture. They have already done their work.

SUMMARY

- During filming, every one or two days the film is sent to the lab for immediate processing. The raw, uncut film is called the rushes.
- The editor and the director view the rushes. From that they determine which scenes are fantastic, which need salvaging, and which are going to be reshot.
- Notes during that meeting are carefully taped. The editor and the assistant editor later match the film, the notes, and the script supervisor's notes. Any irregularities—such as missing shots—are reported to the director.
- Before calling an end to the production, be certain that every shot, every pickup and insert, and every title shot is complete. Principal photography must be completed before letting everyone go. There will be no second chance.

12

Sound and Music

Low-budget picture sound and music is a complicated, perplexing subject that must be solved somehow. Fortunately you have more time to do the sound and music than you had for the principal photography. That does not mean, however, that you can procrastinate. The hidden costs are still nibbling away every day at your total budget. The interest rates on the monies invested in a picture are high and this, added to your regular costs, means time is money.

The first great barrier that you will have to surmont is the dialogue. If you use direct sound, it may be so "dirty" as to be unusable. Stop right where you are at this moment and listen with your eyes closed. Do you hear that radio you didn't notice before? A bird singing and the sound of traffic in the distance? Where did those sounds come from? We rarely think of all these things in our daily life. Our brain is very selective; we hear it all but ignore most of it, paying attention to just what we want to use. What does all this have to do with filmmakers? Plenty. If we opt to shoot on a soundstage, fine, but daily rentals are high, although the sound is of good quality. It will be necessary to rent a blimp for the camera and forego the use of a long zoom lens, or rent a noiseless camera at a much higher price. Then again, direct-sound shooting slows work on the set. Mistakes in diction, enunciation, volume, or intent mean that you have to make more retakes. So what can be done to keep costs down?

First, let's take a look at what the producers in other countries do about dialogue. In Italy, the largest producer of low-budget pictures, direct sound is rarely used. Practically everything is looped. They can hire stars from the United States, Europe, or anywhere. All they have to do is mouth the dialogue in English or Italian, as the case may be. Mexico, another country that makes low-budget pictures, shoots many of their films without direct sound. How do these national industries do it? It's quite simple — a guide track recording is made of every take. When the workprint of the entire picture is edited to the satisfaction of the director and editor, it is used to loop the dialogue.

Foreign players are accustomed to working this way and are quite adept at replaying their own voices as they appear on the dubbing theater screen. They can listen to the guide track of their own original "dirty" sound, and improve their interpretation of it under the guidance of a competent dialogue director. The dialogue can be given more feeling, meaning, and subtleties without changing a word. You may ask, "Is the quality of this manner of dubbing so bad that we don't do it here in the United States?" or "Why don't they use this method in Hollywood?" The answers are simple: They do it in Hollywood every day and, to a greater or lesser degree, in every picture — but they don't tell you about it. Great noise and talk has been made about direct sound, but how do you think they do that dialogue in moving car shots, extreme long shots, and so forth? Direct sound? No, it's postsynced.

Of course, all the soundstage work is direct. Can you tell the difference between one and the other? Not if it's done properly, meaning that each loop should be repeated three or four takes under the supervision of a competent dialogue director. He or she stands at the elbow of the player in the recording studio and corrects the dialogue as the loop flashes by the screen through one or two repeats, then makes 258 take 1, 2, 3, or 4. Since this is done quickly, more takes are possible to get perfection.

The dialogue director moves the player closer or farther from the microphone to get presence, according to the distance of the camera from the player. As each player does his own loops, he can concentrate completely on his task.

Some critics accustomed to seeing "dubbed" versions of foreign pictures are against using this process. Dubbing is another thing entirely. It consists in adapting another language to the lip movement of the original language; it is never completely satisfactory, to put it mildly. Never confuse looping with dubbing. With looping, you can exchange the voice of one actor for another and no one will be the wiser. Another great help to successful looping is the use of the correct sound effects tracks, more of which we will discuss later. Have no fear of looping; your favorite Hollywood star has done it time and again, and you've never perceived any difference in the quality.

There are several ways to loop a picture. The old, timeworn system was to cut out each piece of work print and form it into a loop with a number, using special looping leaders with a moving central line to tell one when to expect the dialogue. Later these loops were cut apart and put back into the complete work print. It was tiresome work for the editor and the editor's assistant.

Nowadays the projector is used on a "rock and roll" method, returning the work print to a point just before the start of the dialogue to play back the recorded voice, or to record a new take if the original was not satisfactory.

However, a brand new system using a video cassette of the work print has come into use and promises to put all former systems out of business in a hurry. This new system can be used two ways: If video tape editing equipment is available from the start, all the negative shot can be transferred to tape to be edited to video cassette work print. Once a satisfactory cut is obtained it must be "frozen," which means no further editing changes should be made.

The definitive tape is then run in sync with a multitrack audio tape. All the effects tracks, music, and dialogue tracks are then transferred to this master tape. When all the tracks are recorded, they must be mixed and transferred to the tape that will be used to make the sound negative. Take note that the paperwork will be the same as with the magnastripe method. After all the mixing is completed, the original negative is cut to conform with the cassette work print.

If your editor wishes, he or she can make up a work print from positive film on the moviola. This method provides the editor and director the advantage of seeing the film projected on a theater-size screen rather than on a TV monitor. After approval, this work print can be transferred to video cassette to follow the above-mentioned path for audio postproduction.

In the following paragraph we will describe the traditional method in detail. It will be noticed that the magnastripe method is labor-intensive and, for no other reason, it will disappear in favor of the video cassette, which eliminates much of the drudgery from editing. This method promises to make obsolete much of the slugging and synchronizing processes, but it doesn't eliminate them entirely; they are replaced by other mixing processes.

Synchronized projectors and recorders are used in the processes, as well as all recorded dialogue, usually transcribed to 35mm magnastripe, to be synchronized on the moviola with the original work print.

Look around the sound lab field to come up with the best choice. Pricing in this field is marvelous to behold, from rock-bottom to astounding gold-platers. Go where the low-budget producers go. Don't spend a lot of money on this — no one will ever know how many channels were used.

After synchronizing every magnasync recording of the dialogue, the editor will make the sync marks on both tracks with the yellow wax pencil before handing the whole reel to his or her assistant to splice together using a CIRO tape splicer and a synchronizer. This should be a five-wheel model to hold the different tracks. The assistant will fill out all the blank spaces with "slugs." The cheapest form of slug consists of rolls of ordinary discarded copies of films that have been cleaned after cutting out all sprocket defects. Any clean 35mm film can be used for slugs; don't waste good magnastripe material on them.

When the splicing of the dialogue track is finished, it should be run through the moviola in sync with the work print to check that every bit of dialogue is in sync. A frame here and there might have to be added or subtracted, but always remember to compensate for this immediately after the off-sync portion, otherwise all the remaining track will be out of sync by the mentioned amount. Just add or subtract a slug after the change.

The paperwork for the looping begins on page one of the script. Beside the first dialogue with number 1, for example, write loop #1 and, on a separate piece of paper, transcribe the dialogue preceded by number 1. If the dialogues are long, divide them into two or more loops. Don't try to make them too long. Short sentences are easier on the artist, and the editor will have less trouble splicing them together so they will sound just like a single speech. It is necessary to put the name of the player after the loop number as each take is recorded. The dialogue director will tell you which one is the best take. This is done right on the spot. Write down T-5 OK and check it off on your own script next to the loop number.

Figure 14 is a page of looping dialogue. It will tell you more than words. A full-length feature can be looped in its entirety in two to three working days. These

Loop	187	Bancroft:	What are you doing Nick? You can't leave—can you?
Loop	188	Nick:	We're not leaving. Just charging the battery a little—
Loop	189	Nick:	Just about dead—
		Kitty:	Hello—
Loop	190	Bancroft:	Well–see you at dinner.
		Nick:	Thank you but we'll have dinner on board today.
Loop	191	Bancroft:	I see—well—see you later.
Loop	192	Kitty:	Why did you refuse, Nick? You know in a little while we are going to run out of food.
Loop	192	Nick:	You know why—and besides I get tired of those damned intellectuals—they're always thinking.
Loop	194	Kitty:	Don't you think too?
		Nick:	Sure I do—but I don't tell anybody about it.

▶ ***Figure 14*** *Sample page of a looping dialogue.*

should be eight-hour recording sessions with a break for lunch of one hour. More time could be spent on looping, but recording studio time is expensive.

About this time music begins to enter the picture. Composing music for the screen is very special. In the first place, a composer of symphonic music is accustomed to the classical diversions or movements stated in a given way: allegro, allegro non troppo, adagio, etc. Although all the classical techniques may be, and often are, used in screen music, the movement as such disappears completely. The structural form has changed radically. The screen composer must be, above all, a gifted arranger. To wit, the basic theme for the heroine may be stated in six different moods and in varying lengths from brief passages to lengthy expositions. Composers with the inventiveness and flexibility for the needs of the big screen are scarce and expensive. When you find one you can afford, you'll note that he or she is overworked and usually under great pressure. Let us explain how this music is done.

The composer must be present when the double-headed projection (image plus dialogue) of the entire picture is run. This will give him or her a feel for the requirements. Be sure to specify what type of music you want — modern, classical, electronic, atonal — and what size orchestra you can afford, and give the composer a script. At this point he or she will make mental or written notes of the things that are pertinent to the music. Later the composer will make an appointment with the editor, and in the editing room the editor will run the picture, reel by reel, on a moviola. With many stops, starts, rewinds, and repeats, the composer, with the aid of a stopwatch and notebook, will time out the places where the music should go (see Figure 15). The director will already have made known whether he or she thinks the work must be heavily scored or not. (Pictures of some years back had music under everything, many times using the benefit of a more pointed score.) The composer will then create themes for the protagonist, antagonist, supporting players, and the main title, and usually a full treatment for the end of the picture. Each bit of music will be numbered and timed with a written description of the action. For example: "#19. Nick hides the box, 2′ 23″. Bancroft looks for the box, 40″." Presently the composer will arrive with a lot of sheets of music and play it for you on the piano. Any changes or suggestions should be made at this time. Some composers will present a choice of themes at the very beginning. Once they have been chosen, stick to them.

Producers are always happy if you can work in a theme song which may turn into a big hit with the public.

Composer's Notes on Music Picture: Deadly Reef			
1.	Neopolitan Theme	Normal Theme Vocal Intrumental-triste Dramatic Funeral	1'30" Orchestra 1'17" Orchestra—Voice 2'00" Orchestra—Organ 20" Piano solo 44" Accordion—Orchestra
2.	Kitty's Theme	Normal Theme Triste Love Scene Nick Vocal (end film)	2'10" Instrumental—Beach 1'10" Instrumental—Yacht 3'09" Piano + Orchestra 2'40" Orchestra + Vocals
3.	Kitty's Dance	Sync to Projection	Joe's Rhythm Group
4.	Theme Love Scene	Sync to Projection	Orchestra
5.	Underwater Music	Submarine	1'30" Instrumental Percussion
6.	Kitty's Theme Variation	Kitty Lonely	1'16" Piano Solo

▶ ***Figure 15*** *This is a small sample of several pages of notes made by the composer after viewing projected work print and timing scenes using a moviola prior to composing music.*

In most texts on screen music, you will see a large orchestra posed in front of a giant screen with the music director on the makeshift podium, and so forth. Stop dreaming. This is a low-budget production: you won't have any click track headphones at all. Your composer will get the musicians together and conduct them. Musicians for this type of work are usually excellent sight readers and, if they have worked for the composer before, know exactly what is wanted.

The composer's fee is subject to negotiation. The musicians' salaries are not; they are specified by union contract. High salaries have driven some low-budget producers abroad for their music, but even that entails travel, contracting abroad, language problems, and so forth. Original music is not cheap, but it's the only way to go. On some modern rock pictures, the group from the picture will do all the music.

There are music companies that can provide prerecorded music for any mood, theme, or necessity. Unfortunately, this kind of music has been used time and again for documentaries, commercials, and you name it. It just won't work for a new motion picture.

Music recording sessions are fascinating experiences. The music for a picture usually can be done in two days. The music director or the composer moves right along, explaining, "Now we'll take #43. This is the oboe solo, followed by a solo of the violin. It is 7 seconds long. All right, ready?" Then, after the 7 seconds are recorded: "Good. Now let's go to #44. This is the theme for the heroine — let's be sure to hear that cello." The sound technician will move a microphone, a couple of tests will be made, and the director will say, "Let's do it." And so it goes, hour after hour. When the recording sessions are finished it will all be down on tape — stereo preferably — just in case you might make a sound track album. Then it will be ready to be transferred to magnastripe for the editor. *Music* editor, you ask? No such luck. On low-budget films, one editor usually cuts the whole works.

Some composers will make a package deal, providing all the music for a fixed figure. If possible, this should be contracted. Just like packaging players with an agent, laboratory services, etc., this arrangement saves a lot of work and provides the same results. Composers sometimes have already written and recorded music for sale, but it very seldom fills the requirements. However, it pays to look into everything. Music being one of the problem areas of low-budget production, it merits study and judicious methods. Please avoid piracy or pirated music — this can lead to very serious trouble and can stop your picture from being distributed and exhibited.

SUMMARY

- The question of dialogue is a difficult compromise in low-budget films. Direct taping on location can create a "dirty" feel to the sound track. Filming the entire movie in a sound studio, however, can be expensive.
- Looping the dialogue, or having the actors dub over their lines, is a cost-effective way of cleaning up the sound. There are several ways to loop dialogue, ranging from labor intensive magnastrips to transfering the workprint to video cassette, editing it, and then transfering the audio track to the master tape.

- After the dialogue is complete, a composer then scores, or adapts classical scores, to the musical needs of the film.
- The composer assembles the musicians and conducts the music.
- Don't overlook the possibility of including at least one tune that could be a potential pop hit. The publicity couldn't hurt.
- Avoid prerecorded music, which is most often highly overused, and steer clear of pirated music.

13

Editing the Film and the Sound Tracks

The mechanics of editing are quite simple. After a brief explanation, anyone can cut off a slate and splice two pieces of film together. That is not editing, but simply a routine, cutting of the slate, marking the numbers on the film, and so forth. An editor runs the piece of film through the moviola, then reverses it, runs it forward a little and back, then makes a mark across it before or after the frames where the cut should go. A highly tuned sense of timing tells him that this is where it should go. It could have been a few frames one way or the other, but here is the cut. Later the editor will run the film sequence by sequence and, sometimes, break apart the splice to take out something, or call for the assistant to look for that little scrap of #492 in the proper case. (Good editors never throw *anything* away until the whole job is finished, and even then all the outtakes and important pieces are in a can on the editor's shelves.) At times you may desperately need just a few frames at the beginning or end of a shot for matching or some other reason. The editor can find them and inset the missing piece.

The editor or assistant should cut out all the slates marking each start, and finish the cut with an X, using a grease pencil (preferably the yellow type used in bowling alleys), annotating the number at the head of the shot thusly: X - 259. He or she should roll this into a tight roll with the number out, put a rubber band around it, and deposit it in its proper cubbyhole or correct place on the rack.

Perhaps you are wondering why we specified a moviola in place of a more modern flatbed editing machine. The reasons are several. First the rental cost is cheaper for the moviola. Second, a flatbed editing machine is more difficult to transport from one place to another. Third, in spite of all the talk going around, a moviola is faster to work with. Of course, the moviola image is smaller, but that is of little concern.

With a bin behind it and a reel horse to the left with a splicer nearby on the rewind bench, the setup lends itself to really rapid cutting. No lifting rolls here or there. Just open the gate, slip in the piece of film and view it. Another bin nearby with pegs for holding sequences helps turn out the work rapidly and efficiently. Synchronizing is very rapid with the magnastripe sound on the left side; just opening or closing the clutch allows for advance or retard, and all marking is right in front of you.

Some may think this method is old-fashioned, but it's the best way to do the job because of its simplicity. Remember, we are not putting on a show; we are cutting a low-budget motion picture, and it has to be done at the same pace the director uses on the set. We personally know of a case where the trailer for a feature film was cut on an old-fashioned black model moviola with a bullseye lens.

The CIRO splicer using Scotch tape is one of the greatest inventions of modern times for helping the editor with this laborious task. Both sides of the work print should be taped, as it will run through the moviola and projector many times before the picture is finished. Magnastripe can only be taped on the back side. Some editors use the diagonal cutter for magnastripe, others the straight one. Theoretically the diagonal cutter avoids clicks.

The editor usually strives to finish and freeze the workprint as soon as possible. The director must be with the editor all this time, supervising the cut. Of course, a good editor needs little supervision, but the timing and sense of a picture must be the director's province. Sometimes entire sequences or shots have to be transported, or even eliminated, and this should be the director's decision. Editing a feature film is an art, as stated before. Of course, it can be learned, but personal sensitivity aids in its fulfillment.

Once the work print is frozen, all the other tracks will be synchronized to it. Be sure it's synchronized properly, as any later change will require a corresponding change in all the other tracks. When the workprint and the dialogue tracks are finished and projected on a double-head projector, the editor can begin to work on the special effects track. He runs the work print from the beginning with notebook in hand, making notes — #12 telephone ring 5″, #13 sound of shower 10″, #14 sound of night bird, and so forth. When the list is finished, the editor can start the search for tapes of these effects. Some sound studios have extensive libraries of effects and can supply magnastripe copies for a price. (Sound effects suppliers offer both tapes and discs. Stick to tapes. Records always sound like records, for some reason.) All these effects are transferred to magnastripe and numbered, then synched on the moviola to match the work print. Be sure to have start marks and beeps on all sound tracks. The sound track is slugged with clean, used copy material just like the dialogue track, using the synchronizer as before.

The music tracks come next. Music and musical effects usually must be faded in and out, and many times one part will overlap another. Usually two tracks are sufficient.

Another track of nonsync effects, such as wind, waves, bird song, traffic, etc., should be made according to the needs of each sequence. These are quite easy to edit and should present few difficulties.

At this point we will discuss one of the most critical and important sound tracks of the entire production. It goes by several names one of them is "Foley." This is a background track, made by specialists, of everything that is not included in the above-mentioned tracks. Practically every large motion picture center has one or more persons who specialize in producing these effects. You'll have to look hard for

TRANSFERS

Transfers — Special Effects Track for the Picture "*Deadly Reef*"

		Feet
1.	Rapidly accelerating motor, cabin cruiser	300
2.	Constant sound-motor, cabin cruiser	400
3.	High speed pass-motor, cabin cruiser	150
4.	Sound of starter & motor, cabin cruiser	250
5.	Cries of seagulls-isolated & clear	250
6.	Sound of large waves breaking	350
7.	Sound of small waves breaking	450
8.	Soft breezes	1,000
9.	Strong winds	400
10.	Rain	300
11.	Thunder and lightning	150
12.	Telephone bell	20
13.	Sounds of shortwave radio	200
14.	Cries of night birds	50
15.	Sharp loud cry of night bird	30
16.	End of record-playback Kitty's dance	100
17.	Sound of shower	150
18.	Sound strong diesel engine-Coast Guard	250
19.	Sound air bubbles-scuba divers	200
20.	Motor light plane-passing overhead	100
21.	Sound of heavy blows-fists	50
22.	Sound of blows with grunts	50
23.	Sound of sharp blows to head	50
24.	Sound of slaps on face	30
25.	Sound of tic-toc pendulum wall clock	250
26.	One person applauding off camera	30
27.	Sound big diesel turning over slowly	100

Transfers to 35mm magnastripe from ¼" tapes, records, or other material must be made in order to make up an effects tract. This track is synchronized and slugged in the same manner as a dialogue track. The footage numbers are noted on the form to be used by the mixer during rerecording. Sync leaders are attached with magnetic "beeps" for synchronous starts. Some effects may have to be recorded especially for a picture if they are available.

one because they don't often advertise. Their charges are usually fairly reasonable. The background sound artists sit through the projection of your work print, reel by reel, taking notes. When they arrive the next day at the sound studio (which must have good soundproof 35mm sync projection), you will be flabbergasted as they unload a small door, a battered typewriter, a wooden platform, some sacks of sand, a baby bathtub, some boards, a shower hose, and all kinds of junk! Don't ask questions — just watch! Taking their notes from the first reel, they will call for the first reel of the work print. They have put their microphones in the right places and will begin the process of recording. Their director gives a marvelous dumb show. While viewing the screen, they will lift and put down a coffee cup, light a cigarette with a match or lighter, walk on the boards, close doors and windows, sit down in chairs, rustle garments and in general create every natural sound necessary to enhance the image. All this is done in absolute silence as any word or unwanted sound will be recorded at the same time. When the 1000′ reel is finished, they will play it back in sync. If it's not exactly right they will do it over again.

Most of these artists of background sound prefer to work by the reel, rather than in bit and pieces. So, in one working day, they will have given you a track that will make your picture as true to life as — and sometimes better than — directly recorded sound. Your editor will run each 1000′ reel of magnasync on the moviola and fine-tune any disparities in sync. *This track is the secret, the **big** secret of low budget sound success!* No ordinary buzz track can compare with it.

By now you will have the following items:

1. a work print
2. a dialogue track
3. a synchronized sound effects track
4. a non-sync effects track
5. music track #1
6. music track #2
7. the "Foley" effects track.

All these tracks are now complete with their sync leaders beginning and end, and their respective "beeps." Although we will talk of titles in the next chapter, let us suppose they, too, have been finished and spliced into the work print.

The editor must now make the entire program for recording all these tracks in the sound studio. Starting with reel one, he will put the headings, as indicated above, with the footage on the left side of the page, indicating when each effect must come in or go out. This must be double-checked, and all the tracks also checked out on the moviola before going into the rerecording session.

Rerecording is one of the key processes in the confection of a motion picture. In the gestation of a film there are many critical elements, and this is one of them. At this point, the entire sound process comes under the hands of one person (or at the most, two). Here is what occurs on this critical day: The work print is mounted in a synchronous projector at the recording studio. All the tracks are mounted on synchronous 35mm playback machines, one for each track, and all are started together with the picture; the mixer sits at the console with the program produced by the editor in front of him or her. By manipulating the controls and pushing buttons, the mixer can alter the volume and consistency of the sound to fit the requirements of the image being projected on the screen.

The first time through is rarely recorded, as the mixer is becoming familiar with the cue sheet and the picture itself, in addition to any peculiarities in the sound track. As every effort was made during the transfers of original material to the magnastripe to maintain the correct level of recording, there still may be some inconsistencies. At the same time the mixer utilizes taste and judgment as to the correct levels to be maintained by any track at a given moment.

The results are recorded on 35mm full coat recording material of the best quality. The mixer will run through the entire reel, then play it back. If not satisfied with it, he or she may call for a rerun of the material, which must all be rewound to the correct start marks.

When the first reel is satisfactory, the rerecording session continues on, reel by reel, until the end of the picture. Usually one full working day is needed for the session. The picture at this point is contained in about 72 1000-foot cans. These cans should all be labeled on top with round stickers and on the sides (around the edge) with felt pen: Producer — Title — Music reel #3, etc. Identify everything from the first day. Your assistant will build up some muscle loading, unloading, and carrying these cans around. Provide cardboard boxes. Leftover boxes from film manufacturers are just the right size — try to get them; you will need quite a few.

At this point it's necessary to consider knowing a bit about rerecording. The international track will be of great aid in making your picture more salable to non-English speaking countries. Some of these countries will accept pictures with subtitles so, before returning to the track, we will talk of subtitles for a little while.

There are two basic methods of getting foreign language subtitles on your picture; the use of each is decided by the number of prints you are going to make. Let us assume that less than ten copies are to be made in French or German. The easiest way is to use a method called TITRA. In this process, the translation of the dialogue is prepared and printed on sheets of glossy paper. These are reduced to miniature size by a photoengraving process to occupy the correct space on the frame. The copy is marked to tell the operator the correct footage for the impression of this small type, which is heated and pressed into the emulsion, burning it off, and literally leaving a transparent subtitle. This process is repeated for as many dialogues as there are contained in the whole picture, starting at the beginning. The cost of the process includes the translation of the dialogue and all the work. It's the easiest and most-used system for subtitles in any language. Also, if you have a large number of used copies on hand, this system is the best.

The second process for subtitles, when a large run is to be made and new copies are called for, consists in making a transparent band of subtitles to be placed in the printer with the negative at the time the copies are to be made. The dialogues are printed and photographed with a motion picture camera, just as if they were titles. The positive print with black letters on transparent background is slugged out with transparent leader using a synchronizer and running a copy on the moviola to locate the dialogues. This system is best if a large number of copies is needed for special markets.

There are markets, however, that refuse subtitled pictures and genuinely prefer to hear their own language in a dubbed version with all its deficiencies. Countries like Spain, Italy, and France come to mind, although they are not the only ones, by any means. The usual process is simply to make up a complete new sound track with new effects, new dialogue, and tapes of the original music. This new track usually sounds low in quality to the person who has heard the original track, but the public of those countries accept it as they don't know what the original track was, anyway. Of course, this method works against the quality of the picture and its final acceptance.

Here is where the international track comes into place. As the rerecording proceeds, all the tracks except the dialogue are fed simultaneously to another synchronous recorder; this track is sometimes known as a minus dialogue track, and usually it's done on 1/4-inch tape, in spite of the stretch that may occur. A copy of this tape is provided to the foreign distributor, who makes a magnastripe copy. This is reviewed on a moviola with a copy of the picture, and any minor sync changes are incorporated. The distributor then adds the second language.

Now that you have the work print and the full coat sound, you are ready to return to the laboratory. In Chapter 15 we will follow the course of the material as it winds it way along the path of production.

While on the topics of editing and synchronizing the sound, we should not forget the trailer. How could we? Some moviegoers state categorically that the trailer was the best part of the movie — and so it is.

An excellent trailer will bring the cash customers to the box office; a bad one will kill interest. There is a good rule on trailers: as an old stripper used to say, "Show enough to tantalize, but never give away too much." A good trailer should not be too long. Some of them run for almost ten minutes and, after seeing them, you're sure that you saw the whole picture. Maybe you did! Keep it short and concise and lay it on strong. By now the editor will have called out delicious bits of outtakes or even ordered dupe negatives of some portions, but usually that won't be necessary.

Start the trailer with a "hook"—something that will fix the patrons in their seats when they are about to go for popcorn. Keep it coming for three to five minutes, with repetition of the title at the end. The trailer is made just like the picture, sound tracks and all, except dialogue bits are short, or it may be narrated. A good trailer is the equivalent of the come-on at a carnival sideshow.

Use part of the theme music already recorded, you won't need all the effects and multiple music tracks for this effort. Do your rerecording sessions right after finishing the feature. Never use opticals or any fancy elements on a trailer.

SUMMARY

- The mechanics of editing are simple; the art and sense of timing are learned through experience.

- An editor runs the work print through a moviola, reverses it, forwards it, reverses again, then marks just before or after the frame where the cut will go. The film is then cut, spliced, run through the moviola, recut, and respliced.
- Once the work print is frozen, all the other tracks will be synchronized to it.
- The dialogue track will be laid first, then the synchronized sound effects, two musical effects, nonsynched effect, and, finally, the Foley effects.
- The tracks are then rerecorded and mixed to the proper sound levels.
- There are two processes to creating subtitles for the film. One actually burns the tiny titles onto the film, creating transparent spaces. The other involves creating a transparent band of subtitles to be placed in the printer with the negative at the time the copies are to be made.
- Now is the time to develop a great trailer. A good one will generate interest and a bad one will kill it.

14

How to Make the Main Titles

Motion pictures have used every type of title treatment imaginable, from stop motion, animated cartoons, paintings, engravings, and etchings to printed titles and animals in every shape, size, and variety. Many times the titles are made by specialists. Some are so good you may remember them and forget the picture.

The main titles of a low-budget picture should be simple, straightforward, and easy to read. However, there are some points to consider. They should be professional titles. Even though production money at this stage can be scarce, there are different ways of handling the titles. Commercial titles kits are worse than useless; they stamp your production as a home movie from the beginning. Well, then, what are we to do?

First, study the books on titles, then go out and make your own. One popular producer sticks to standard procedures. He has all the titles set in black Letraset adhesive letters and white poster board. When they are all made up, he takes them to a photolith plant and has them converted to photolith negatives (with all the pinholes touched up). Then he places the titles on a light box and photographs them, using an old Mitchell Standard camera with fade-out device, using high-contrast negative BW stock. This negative is then slugged out with clear leader to the length desired and the work copy is used to size it. Then he has an interpositive made on Kodak's marvelous 5243 film, and has the lab run these two through the printer together, using 5247 material for a dupe negative, giving him white letters over the original moving background. This procedure turns out very professional moving background titles for quite a small amount of money. That's just one way of doing them; there are many others.

Optical printer houses specializing in titles do excellent jobs, but are a little high in cost for your operation. Another producer likes to use striking, poster-type paintings as a background. He orders the letters, painted in white ink on transparent overlaps, and photographs them in register over the painting. Naturally it works very well when done properly. There is one recommendation we would make: Don't use red letters; this color glitters and is difficult to read. In the old days, red was very popular on color epics. If possible, avoid the use of opticals; they tend to be very expensive in the long run. In Hollywood you can find several economical title houses besides the big popular ones.

In the same way, artwork can be done by any good competent poster artist. There are many of these artists from coast to coast who do this work. They can really surprise you with the quality of their work.

Great care must be exercised to avoid a "jumpy" title. The best cameras, other than the very special ones used by title houses, are the Mitchell Standard or the Bell and Howell with register pin movements. Both of these cameras were used for motion picture photography until replaced by newer models, lighter in weight and soundproof. But these old-timers are almost infallible and make very good title cameras. If you are going to shoot your own titles, rent, don't buy; but your best bet is still a bargain title house—they do this work every day.

SUMMARY

- The main titles of a low-budget feature should be simple, straightforward, and easy to read.
- Keep them professional.
- Avoid jumpy titles. The best, sturdiest cameras to use are the Mitchell Standard or the Bell and Howell with register pin movements.

15

The Answer Print

With the rerecording terminated, we return to the laboratory with the work print of the feature and the full coat sound recording. At the same time, we deliver the work print of the trailer and its sound track.

It pays to leave all the negative cutting to the laboratory, as they have highly trained personnel and special splicers, plus a completely dust-free area to carry out the job; usually this area is pressurized with filtered air to prevent entry of dust when a door is opened to the outside. Here each piece of negative corresponding to its counterpart will be searched for and carefully spliced into place on a special hot splicer. Then the whole negative is double-checked to insure that it is in complete accordance with the work print.

Meanwhile, in another part of the lab, the 35mm full coat recording will be transferred to 35mm optical negative materal and developed. When finished, this is usually returned to the negative cutter, who checks the first few feet of sound track on a clean moviola to be sure that the correct sound track negative is mated to the correct image negative.

Then the negative is turned over to the timer. The timer's level of expertise will determine, to a great degree, the final quality of what you will see on the screen. With the aid of a Kodak or Hazeltine color video apparatus, the timer determines the density and the amount of the colors to present in each separate piece of film, while looking at a color monitor and adjusting four principal controls until the image is satisfactory. The director must run the work print before this work takes place and indicate the scenes that need correction. This is particularly true of day-for-night work, for, if printed too light, it will look just like day and not night! The information taken from the video apparatus is used to set the controls that punch holes on a control tape, which will automatically make the pertinent changes as the negative runs through the additive color printer.

The laboratory can and will, at this point, make very large changes in color or density if they are so desired. Scenes can be darkened to almost pitch black or lightened considerably. However, it must be kept in mind that there are inherent limitations, due to the original color balance and exposure of the negative. The timer can help correct on deviation from the normal to a certain degree, but no timer can produce miracles at this late stage by creating something that did not exist in the original; he or she can only modify!

Do not expect absolute perfection in the answer prints. The timer will make notes as to pertinent changes in some scenes, and repunch the tape so that subsequent copies will be closer to the ideal print.

A note to the parsimonious: Reels from the answer print are not discarded. The defective scene will be excised and replaced with a new piece, and this reel will go out to be exhibited as would any other. The laboratory will make reasonable changes, but will limit them to avoid losing time and money on the perfectionist who may never be content with the results.

The answer print, when finished, means that the picture itself is finished, but that does not mean that you are finished with it or it with you. This is just the beginning of another vital phase in low-budget feature film making. Preproduction, production, and postproduction are finished, and now the results will be there for all to see.

The producer and friends are dying to see the results of this latest venture. Some will praise; others will criticize and blame all defects on the director or the producer, if they are that type of friends. Take courage—but you will have that if you've gone this far—and set up the private showing. This type of affair is prevalent, so accept it for what it is. Take your own claque to back you up and throw yourself into the fray. If your company has a publicity person, invite him. If you have friends in magazines or papers, get them in. Remember, a word publicity, even if not good, is publicity all the same!

After the showing, there is a polite sound of applause as the producer's friends slap him on the back and say, "Well, you've done it again!" or something similar that leads to dubious and sundry interpretations. Everyone goes off for a late snack or drink at some restaurant. You, as the director, begin to think of the next great picture you'd like to do. (A director can be a real glutton for punishment! But we wouldn't have it any other way.)

SUMMARY

- Deliver the work printer, the trailer, and its sound track to the lab. Leave all the negative cutting to the laboratory as they have highly trained personnel, special splicers, and a dust-free environment.
- Each piece of negative corresponding to its counterpart on the work print will be carefully spliced into place on a special hot splicer.
- A 35mm full coat recording will be transferred to 35mm optical negative material and developed.
- The timer adjusts the density and color balance.
- The answer print at this point isn't absolutely perfect. It can and will be adjusted.
- Now is the time to set up a screening with the producer, publicity people, and friends.

16

Publicity and Festivals

Now that the picture is finished, the hard work begins: What do you *do* with the picture? Every picture requires publicity materials. We have already told you how to get the necessary stills. This publicity work starts right away.

The same artist who did your title paintings can do your posters, given photographs and a text to work from. Your printer will want material for the press book and ask for your ideas to make the layout.

The *press book* is not a book and has nothing to do with the press. Some press books are quite simple and some are extravagant, exotic, and beautiful, not to mention expensive. The press book represents probably the best of all the expenditures that you will make in publicity material. The press book is your calling card to distributors, the press, or magazines. Wherever you go, it carries all the information you need to tell everyone about the picture.

A good press book should be at least letter-size when folded once, making it into four pages. The first page should be almost a poster (see Figure 16). The next one must take care of all the screen credits, the third a brief "come-on" synopsis, and the fourth some action and attraction photos. Large press books put out by some film companies end up in the trash because they are too large to store anywhere. If they are really good, your distributors will hang them on their walls. Make every effort to prepare the best press book possible. Make an elaborate list of all the distributors in the territories where you would like to do business, and send them a copy with a cover letter. Many sales have been made from just such publicity.

The large theater poster is disappearing, sad to say. Many from the past have become collector's items. They are costly to produce, but your distributor may require them. The original artwork can serve a dual purpose if it is used both as one of the pages of the press book and as a quarter-size reduction, a popular item since exhibitors have less wall space today than before, especially in the new multi-cinemas.

Many producers make up six different small posters, each one bearing an eye-catching still from the picture. Because they are smaller, the exhibitors will put

▶ ***Figure 16*** *A large, full-color poster using scenes from the production in Spanish was prepared for the Latin American market to attract distributors, exhibitors, and the viewing public to the picture.*

them up in theater foyers, or whatever place they think will be better for the publicity, if the posters arrive at the same time as the trailer.

Publicity cuts of advertising, screened for printing and ready for paste-up, are appreciated by exhibitors and distributors that run press campaigns in local newspapers. Do them in several sizes so the users can make a choice depending on their purchase of one or two columns of horizontal or vertical publicity.

A special hot number at markets and festivals is printed postcards with the title and an attractive still on the front and information in the back. Be sure to print these, as this printing is much cheaper than that of photo postcards in the long run. If you have a picture that is susceptible to mass advertising at rock concerts or arena shows, for example, make flyers (one color) on newsprint to shower down on the crowds.

Why go to all this work to do the publicity for the picture? There is one very good reason. If you don't do the publicity, the distributor will call his favorite agency and do it for you, charging off the already high agency prices plus his commission on the income of your picture.

Publicity during shooting is a double-edged sword. If it's only local it won't do your picture any good, and may bring crowds that will almost bring shooting to a halt, in some cases, unless you have security guards to rope off areas in the shot. National and international publicity is always good at this time but, remember, your picture may not be released for many months and, sometimes, for more than a year. If you can get a big release, all the media affordable should be contacted: magazines, newspapers, even TV, although this is generally out of the question for low-budget producers unless it is a real hot property. Then all the publicity media will chase the producer. All the plugs for big pictures that you see on national TV are handled by the big agencies, which really charge big fees.

Make the publicity packages of material for your distributors yourself. This will help you keep the costs down while, at the same time, giving them something to work with.

Festivals! Festivals! The great illusion of all neophyte filmmakers. There are small festivals, big festivals, out-of-the-way festivals, and so on. The calendar is filled with them, the whole year, from January to December. What do these affairs really do for a low-budget feature film? It depends. If you belong to a particular political group, you will receive gratuitous publicity, and possibly be hailed as the filmmaker of the year. You and your film will be invited to many festivals, and you just may make some money distributing the picture through the university circuits that pay a flat fee to exhibit your picture. It's not a very profitable road to take, but personally you'll get a lot of press in the cult magazines and the critics will love you.

However most people who write, produce, and direct low-budget feature films fail to fall into this category. You'll just have to work harder to really earn your rewards. Does this mean that festivals are off the calendar for low-budget films? Not by a long shot!

If you study the festival calendar, you'll find that some of them have side bar film markets. You will, of course, already have heard of Cannes—all the big independents go there. Do you know that about 600 pictures are displayed at that festival every year? It's like Times Square on New Year's Eve. Who is going to pay any attention to a low-budget producer in that mob?

Be sure that the festival you attend is visited by real, honest-to-goodness *buyers*. If it fits the bill, go—not to the festival with tuxedos, champagne, fireworks, and jury prizes, but to the film market. You'll find all your friendly competitors there, plus all the magazines and specialized press, and you'll meet distributors from all over the world. Set up your booth with publicity on the walls, get a video cassette player, and run your trailer 'til you wear it out.

By this time, you should have a video cassette of your complete picture with the trailer at the beginning of the cassette. Don't give out stills to the freeloaders collecting publicity; otherwise you won't have anything left after two days of the event. If you feel you must give something out, use postcards or fliers.

You will, no doubt, enjoy the market. Your film will be projected a couple of times with people running in and out every five minutes. Pay no attention to them; they are just lookers. Invite the buyers and hope they will come. Of course, you won't see much of the festival films unless you stop work on your presentations—which you shouldn't do. You are there to *sell*. The critics will all be there. They won't even see your film, but consider yourself lucky.

Why do independent filmmakers go to festivals anyway? The simple, basic reason is that festivals and markets are the places where they can afford to show their product to some distributors and potential buyers. The majors don't need festivals or markets. Their in-house distribution outlets control all the large theater chains that must take their pictures, regardless of terms and in spite of all the antitrust laws that exist. They are not actually coerced, because in the end it's good for them.

The low-budget producers must fight their battles on any ground available, and this is one way to show their products. Frankly we think you'll enjoy taking your picture to the film market, even if you do miss the festival.

One last word: The best film market for independents is MIFED held in Milano, Italy, every October. It's strictly for buying and selling and, although the cost of attending has been climbing every year, it's still the independent's prime showcase.

The American Film Market in Los Angles is also available. There are two Canadian festivals with markets: Toronto and Montreal. Many smaller festivals also exist in other countries.

Get to those markets if you can. Showing your film will help both the picture and your ego. Do everything you can to market your film. No effort is too small. Work hard at it, keep going, keep pushing the picture everywhere and every way you can. You made it—now sell it! In the next chapter we'll tell you more about how to do just that.

SUMMARY

- The press book carries all the information on the movie including a poster, all screen credits, a synopsis, and some action and attraction photos.

- Don't be afraid to attend festivals, but be certain that the festivals you do attend are not a waste of your time. Huge film festivals will swallow your film, while obscure festivals will not draw the potential buyers and distributors.

17

Distribution

How can you help guarantee distribution before you start? The first way is in the selection of subject matter. If everyone wants a good spy thriller and the market is not already glutted, make a super agent chiller.

A great many independents follow trends, sometimes to the bitter end, when they must face the terrible truth that they are left holding the bag with a picture they can't get rid of anywhere. One way to help your picture is to cast your stars in a manner that will almost surely be accepted in at least some territories. This way you will be able to have a base to work from, in any case. We personally think that the material should be relatively fresh.

There is an old saying that you should not make a picture if you don't have the distribution deal set up. This is completely false. Very rarely are distribution contracts even mentioned before an independent picture is finished. Sometimes a distributor will say "Bring me your next picture, too." But that doesn't mean that he will take it on for distribution. It simply means he will look at it. He may or may not like your picture.

In the United States there are national distributors that work through sub-distributors for many territories, and an agreement with one of these companies will give you national coverage. However, some independent producers themselves go directly to the territorial distributors. Some independents have more than 20 of these distributors lined up for their product. How is this done? By travel and hard work.

Arranging for distribution is a very special part of the production office work. It requires use of the telephone, secretarial assistance, appointments, airline flights, and so forth. Who are these territorial distributors and how do you locate them? Easy, just subscribe to *Box Office* magazine and you will learn who handles what distributor in every territory. By assiduous study you'll get the whole panorama on distribution!

There are also international distributors that will take independent production for overseas territories. How can you tell when distributors are honest or competent? If they ask for money up front for the distribution of your picture, they are not dependable. A good distributor should give you an advance or pay a fixed sum for the territory.

When dealing with contracts, each distributor uses a different format, but essentially all are the same. One thing is absolutely vital—read all the fine print and do not authorize unlimited advertising or agree to share costs, or you may find yourself owing money in place of receiving it. One case comes to mind that ended in a lawsuit. An international hit picture was turned over to a distributor and went into exhibition. It grossed many millions of dollars, appearing on *Variety*'s charts for many weeks. When the settlement came about, the producer was very angry. His share came to less than $100,000.

Eternal vigilance means more returns. If you have a lawyer who knows show business, have him or her read your contract carefully. He or she will know where the pitfalls are and can help both you and the distributor. Get a good deal to eliminate settlements made in court.

A quite important matter to be considered when distributing a picture abroad is the elaboration of dupe negatives. There are several methods for accomplishing this very necessary step. As you will remember, in Chapter 12 we talked of the international sound track. The dupe negative is the counterpart.

The first method is the one used by "pirates" of motion pictures. Pirates are ever-present; practically every picture has been pirated at some time or other. The method used, whether for legitimate purposes or not, consists of simply obtaining a copy of the picture running it, preferably, through a liquid-gate printer, and copying it onto 5247 negative material or its equivalent. Scene-by-scene color correction can be used, just as in an original printing. The negative material has usually been pre-flashed, using the "black box" method to avoid excessive contrast. The negative material is exposed in its entirety with a small, well-calculated amount of light to overcome the initial inertia of the light-sensitive particles. The process is a tightly guarded secret which few laboratory technicians will talk about, although it is widely practiced. Some "pirates" even go as far as making a separate first-class sound negative.

The second and better method, though a more expensive one, consists in making a good interpositive on special material, such as EK 5243, which gives perfect contrast and is so good that a dupe negative made by this process looks just as good as or better than the original negative.

A third process, also expensive, is one of the new favorites: the use of color reversal internegative material. This process gives a negative directly from the negative. The quality is superb and provides a duplicate of the original negative to send overseas for release. This is a favorite method of the large producers.

Some foreign labs will suggest that you lend them your original negative to run off copies. Don't do it. Scratches at best and losses at worst can result. Keep your original negative in the lab vault, where it belongs.

Remember that any or all of this work should be paid for by the purchasers of the rights for the territory involved, apart from the payments for the use of the film in said territories. Provide anything they need—for a price—and stipulate everything in your contract.

Four-walling is one of the great old ways of getting a show on the road. Let's suppose, for a moment, that you have back-to-back two quite interesting pictures of the type known as family fare: the dramatic life of a charming older man living with nature on a tropical island and the adventures of a dog in the Canadian Rockies. Rather than simply putting them out for distribution, you decide to take a couple of years off yourself from other duties, or to employ someone to do it for a fixed monthly salary. You turn into a real old time medicine man.

Let's say you arrive in San Antonio, Texas, and set up in a motel. You locate a movie house, independent, of course, that will rent you the theater for a fixed sum or percentage, as the case may be, for one week. Then you go to the newspapers with your advertising cuts and contract space. You visit the local TV station with your trailers, the local radio station with prerecorded plugs, and maybe have an interview. If you are a club person you visit the local clubs with your advertising, then plaster posters on telephone poles and put stand posters in store windows. You exploit every form of local advertising schools, churches, etc. Ten days or 2 weeks later you pack them into the theater. You stand by the ticket taker at the box office and count the house. When you finish the run, you divide the money and are on your way to the next stop. It's a great vagabond life and, believe it or not, there has been more than one such entrepreneur who has made good money and good pictures for many years.

Another form of four-walling is carried out almost all the time. An independent low-budget producer, in effort to show that his picture has "*legs,*" will rent a big city theater and put on an ad campaign for the showing. The object is not to make any money at the bottom line, but to show potential buyers and distributors that his picture has pulling power, and that in 3 days in New York or Cincinnati it grossed so many thousands of dollars. This is a special form of advertising that may pay off.

The money spent in four-walling of this type would be better spent in other areas on the same picture, or developing a new property for the company. These outings of a picture are generally publicized in the trade papers. Study the whole expense, including the postexhibition advertising, to calculate the entire cost before doing this kind of promotion for your picture.

The international territories are very important to the low-budget filmmaker. The usual divisions are as follows:

1. United States and Canada
2. Mexico, Central and South America
3. Europe
4. The Middle East
5. The Far East, which usually includes Australia and New Zealand.

As a purveyor of a low-budget product, you have a card index with the names of distributors and buyers for each of these territories. By reading the trade journals and going to markets and studying directories you will build up your own reference list of who's who in the distribution world. Such a list will serve you well.

Advances are the key to success. Always try to get an advance. If the stars in your picture are known and their performance acceptable, you have a good box office potential and you can demand and get advances. The reason for this demand is simple: These may be the only monies you'll ever see from a given territory. After some time you will also know just about what the maximum and minimum going prices are for each territory. Outright sales for territories are a much better deal than advances. The preoportion is really the same, but if the contract is finalized, there is nothing more to worry about. This route is highly recommended by practically everybody. Of course, if your picture is a big hit, then you are out of the profit picture. Well, you can't have it both ways.

The story of the fairy godmother comes to mind when we talk of pickups. From time to time we read in the trade papers that Mogul pictures, the big major, made a pickup deal with so and so for various millions of dollars. The majors can and do make pickup deals every year.

Why is this done? After looking over their commitments to exhibitors, and taking into account their production schedule, the head office finds that they will need a couple of pictures to fill out the calendar.

It's becoming more and more a habit nowadays to let someone else's money work for you and take the risk. The upper crust members of the independent producers, in an attempt to break through to the marketplace, have contracted more name players and spent a little more on production; with these efforts they have a fairly good picture in hand. The major with this knowledge goes to the film markets looking for just such pickups. Then the haggling begins. The major has the advantage of course—cash in hand, which is a hard weapon to combat. Naturally the independent producers will not sell for less than the total accumulated value of the picture—plus whatever profit they feel they can make from the distribution deals.

Finally a deal is struck. The independents sigh in relief and congratulate themselves, strolling out to the beach to find a new star for the next great picture. Maybe they did not get as much as they should have, but they're still solidly in business and have saved a lot of hard work. The major has made a good deal and is sure the board will find it satisfactory. The picture is then "cleaned up," that is, any union or other obligations are paid, which may in some cases be avoided because the picture was given a certificate of origin as having been made abroad. If such is the case, it goes through their normal channels of distribution.

Outright sales of pictures are usually just a recoupment of investment by someone who wants out or who, simply by force of circumstance, must get rid of the product. These deals occur often and, usually, there is very little profit in them. Sometimes a company will sell its entire library of pictures for a large sum. Recently a quite large Latin American distributor sold everything to a newcomer, including a library of over 600 pictures.

Just before the great breakthrough of television, an aging star bought the rights to all his films from a low-budget producer, who was all too happy to get rid of them very cheaply. In a few years, with the advent of television, the actor showed his pictures on the TV screen and, when he died, left a fortune to his heirs. Pictures change hands every day. Sometimes the deal is profitable but any picture, however old or new, is a property and has real value. Never forget it.

Producer's representatives are a species of intermediaries that exists only in the world of the low-budget producer. The majors have distribution offices located in practically every part of the world. Sometimes these are affiliated companies representing one or more of the majors, thereby avoiding duplication of effort. For this reason, they have no need for representatives. The small company has neither the time nor the money to properly sell it's picture abroad. This is when the producer's representative comes on stage.

The best representatives are those who really know where and how to sell a picture and who are quite choosy about the products that they represent. This type of company is expensive but usually gets results. After talking with them you'll feel that you've been in the office of a major, where, by the way, the deal may end because these representatives do come up with pickup deals. They usually circulate in the circles of the top executives in the industry, most of them having been there before. There aren't many of these persons.

The other end of the scale is represented by people who want to take advantage of low-budget independent producers, and who may ask for all kinds of money to take the picture to markets. They may promise you everything but have probably never been through the gates of a major studio. Take care and steer clear of these persons. It's always easier to say no from the beginning than to try to extricate yourself from an unfortunate arrangement.

The producer's representative whom you select may not be perfect, but who is? He or she must be a solid business person who knows the overseas market and attends several of them a year to personally meet the buyers and distributors. This person will charge you the going rate plus some expenses, but never up-front money. The deal when made may not be the best, but a good representative will make every effort to properly sell your picture. You will have to look carefully for this paragon of virtue, but there are a few around.

Now a few words about the marketplace. The low-budget producers occupy about 10 percent of the total market for pictures. The majors have solidly 90 percent of the entire world market. You will have to fight hard to get and hold a small portion of this market. This knowledge should serve to reinforce your desire to triumph, rather than encourage you to quit. There are many producers, directors, writers, cameramen, and technicians who work and live out their lives in this field. You must not expect all the largesse of riches, but you should do the best that you can to achieve success.

An independent filmmaker has great advantages, and more than a few limitations as well. Anyone can make a picture with $30 million. But it takes courage, knowledge, and sheer guts to make a feature film from the bottom of the heap. When it's all done, you'll know who you are and what you want to do. The independent will always have precisely what the word means—independence.

Television sales are an important source of income today. The voracious maw of television consumes material of all sorts at an exorbitant rate. Like some ancient god of old, it must be fed constantly with sacrifices, fortunately for the producers of low-budget films and majors alike. Yet sales to television are not as easy as they would seem. The reason is simply that the individual producer doesn't have sufficient available product to satisfy the demands. In addition, individual access to the large television chains is quite difficult, hence the necessity to retain the services of an intermediary, an agent who will put together a package of films to offer to these companies.

Another factor to be taken into account at the present and, as it looks now, in the foreseeable future, is the system of playing off of the majors. You will have noticed by now the diminishing number of neighborhood theaters. Every available screen seems to be first-run. There is now no opportunity for the viewers to see the pictures they missed last month or year, due to the procedure of the major producers that covers from 600 to 1500 screens nationwide simultaneously. This method takes full advantage of massive, one-time publicity of television, radio, newspapers, and magazines, and uses their preestablished evaluation of the picture. The producers can thus extract advances and playdates from the exhibitors. From the exhibition the picture passes directly to television and from there to the video cassette market. Little effort is made to continue distribution to independent theaters not controlled by the majors, as was the custom of the past.

The close relationship that exists between the television networks and major film producers is well known; thus the majors' products come on first. However, there are many TV stations in need of product that may not be able to afford the network prices. These are the target market for the producer's representative.

A booming new market that really promises to have a good future is the business of video cassettes. In spite of wholesale piracy, the video cassette business holds great promise for the independent producer, maybe not as a major source of income but as a final effort at sales to help the total gross of the picture. Don't despise this market and try to extract additional income from the sale. Of course entrepreneurs of this business are like those of any other business; they want to drive down the price as much as possible to make more profit. Oddly enough, if independent low-budget producers did not make pictures, these companies would have less product to sell.

An important thing to remember is that a picture has a continuing value. Witness the fact that some low-budget films have become "cult" films and continue to collect rentals, long after the life expectancy of commercial remuneration has ended. Never consider that a picture is exhausted until every last territory has been sold and every avenue of exploitation tapped.

SUMMARY

- There are international, national, and territorial distributors. It is your job to seek these companies out and to interest them in your film.
- Do not pay money up front for the distribution of your film. The distributors then have no incentive to market your film if they have already been paid.
- Read all the fine print in the contracts. Do not authorize unlimited advertising or agree to share costs.
- When creating dupe negatives, don't lend anybody your original negative.

- It is also possible to take your movie on the road, create your own audience, take the tickets at the door, and show your movie one town at a time.
- Don't neglect the international market, which is very important to the low-budget feature producer. Know who the distributors are for each of the international territories.
- Advances are the key to success in low-budget features. Sometimes an advance is the only money you'll ever see.
- Pickup deals, or the movie version of the fairy godmother story, do happen.
- Television sales are an important source of income, while the major net-workscan be difficult to get into. A better bet may be the independent television stations who can't afford the fees the majors charge.

18

Summing Up

An epilogue always looks in two directions. The first is toward the results of the past and the sum of all the integers to date. It's the bottom line, not in dollars and cents, but in achievement—achievement as the result of hard work and sweat. Many people today look down on sweat, but, by the Gods, how Praxiteles did sweat as he sculpted those marble statues that we still contemplate today.

An epilogue looks secondly toward the future. As one work is finished, another must begin. And to that beginning, to that expectation, we all look forward. All the work and study of the past, all the experiences serve as but a base on which to construct future edifices. When we finish a picture we start over again at the bottom, but not at zero. The sum of all things past are the pavement on which our feet rest. There is only one way to go and that is up!

One of the most important words—and remember, words are simply a means of expressing a fact, concrete or abstract—is "*professional.*" Professionalism is an attitude toward the work you do. A title or a piece of paper conferred, does not mean you are a professional. Professional is a quality, a qualifying adjective that you *earn* as a result of the manner in which you carry out your work. Of course it is abstract, and any abstract quality is difficult to define. However, when you watch a person work, you can tell, feel, or sense that he or she is exercising that quality. In spite of the monetary or pecuniary considerations, and whether the results are crowned with monetary gain or not has nothing to do with professionalism. Some amateurs are more professional in their attitudes and results than many of the so-called professionals.

The true sense of the whole matter is how a subject is treated, how it is approached, and above all, the capacity complete, terminate, or successfully end a difficult task undertaken with the will to triumph—that will to win in spite of all odds. This may sound inspirational in nature. It is, and is sincerely meant to be. There is nothing wrong with ambition, which is the desire to reach a goal and the willingness to pursue that end by all legitimate means.

Do not expect fame or fortune to fall upon you, as it were, from heaven. If it does come as the result of efforts well directed, accept it as such. Remember, in the peculiar field of motion picture production, success and money come and go; the person remains the same. Was Rembrandt less of an artist or a man because he failed to pay for his house?

One famous producer was asked, after a disastrous adventure in show business, "Now that you are poor, what are you going to do?" He snapped right back, "I've been broke many times, but I've never been poor!" This anecdote may have been invented by a reporter—we'll never know. But we do know the inherent truth: Poverty is of the spirit. May our spirit lead us to completed works, whether they be remunerated exceedingly well or not.

It is also hoped that the material in small volume will serve as a guide for future efforts. While it is not meant to take the place of separate texts on each basic subject treated, it is meant to bring together in one volume all the elements of a low-budget production as seen from the point of a writer, producer, director and all-around filmmaker. Some of the material will be recognized as standard procedure and, thus, be well known to all. Other points, we hope, will be novel.

The accumulation of knowledge on any given subject is like a pyramid. The broad base is available to all and the acquisition becomes more difficult and more expensive to achieve as we approach the apex where even a small piece is worth the tremendous effort expanded. To reach the apex should be the aspired goal, although I doubt anyone has ever truly reached that point of infinity and absolute knowledge. After all, we are still humans with limitations, striving for a perfection we may never reach. But we must continue to strive—to settle for less is to surrender.

Where do we go from here? To make our next picture, of course. To learn something new every day. To express that in our work. The fruits of our labor will be there for all to see: the fulfillment of art, talent, work, and study in a finished picture up on the screen for all to see, admire, criticize, and, we hope, enjoy.

May the gods look down favorably on you.

Glossary

Above-the-line costs Expenses pertaining to producer, story, director, and cast.

Advance A partial payment against future rentals in a specified territory.

Angles The angle from which the camera views the scene.

Answer print A complete print with both image and sound track, ready for projection but without final color corrections.

Below-the-line costs All other expenses for technicians, laboratory, editing, film, sound, hotels, etc.

Bits of business Actions or activities of actors within the scene that may or may not be related to the principal action.

Blimp An enclosure that covers the camera to eliminate any sound of motor or gears.

Blocking out the action The director, either beforehand or on the set, determines the movement of the actors and the camera that will take place during the shot.

Breakdown A listing of geographical locations enumerating the shots, cast, action, and calculated time for filming.

Buzz track A nonsynchronized sound track of incidental noises taken at the location of filming, to be used as background sound during a scene.

Call sheets Notices to the actors and crew indicating the work to be performed on the next day.

Camera report A report of negative filmed with slate numbers, footage, and other pertinent information for laboratory and editor.

Certificate of origin The "birth certificate" of a picture, usually a notarized statement as to the country of origin of a picture.

Chalk marks Discrete chalk marks on the floor to indicate where the actor should end movement made on the set.

Changing bag A lightproof bag, used for loading or unloading magazines with film.

Cinematic grammar The shots, angles, camera movements, action within the shot, sound, and music make up a grammar of film language.

Click track A special sound track that, like a metronome, produce clicks at regular intervals, used to time scenes with music.

Close-up shot A shot displaying detail within the scene, such as the face of an actor.

Color-compensating filters Filters used on the camera to adjust the color temperature of the light to the requirements of the film stock in use.

Color gels Sheets of colored plastic material which modify the color of light.

Color temperature A measure taken in degrees Kelvin of light as the film sees it, rather than as the human eye sees it.

Contrast filters Filters that modify contrast and, at the same time, produce a softening effect to the image.

Crane shot A shot made with the camera mounted on a crane, which affords an overall point of view.

Cross fades One scene gradually blends into another.

Cross plot A plot of characters in a screenplay, as related to locations.

Cut The act of cutting pieces of film from a larger complete take, to be spliced together to form a sequence.

Cutaway Cut to another actor or action occurring on a set at the same time.

Cutting in camera Filming different parts of a sequence separately to be edited together, rather than as a master shot.

Day-for-night Shots taken and printed in such a way as to simulate darkness.

Deferments When an actor accepts a postponed payment of a part or all of total salary.

Depth of focus When both foreground and background are in focus at the same time.

Detailed budget A complete cost analysis of the entire production of a picture.

Direct cut Cut to next numerical take.

Distribution deal A contract to distribute the film to exhibitors in specified territories.

Dolly A shot taken from a camera mounted on rails to obtain backward or forward movement.

Dolly in The camera, mounted on a movable platform (or dolly), moves closer to action.

Dolly out The camera, as above, moves away from action.

Dolly shot A shot in which the size of the image is changed by the movement of the camera.

Double head projector A film projector with facility for projecting separately and simultaneously positive image track and sound track.

Dubbing (dubbed) The substitution of dialogue translated to another language.

Dupe negatives Copies of original negative to be used in producing positives for distribution.

Establishing shot Presents to the viewers a scene containing the persons or objects in relation to each other.

Extra long shot Usually covers a very large area; human figures, if seen at all, are very small.

Fade out The scene begins in normal lighting to end in darkness.

Fade in The scene begins in darkness to end in normal lighting.

Filters Economical leaflets advertising a film.

Flashback A complete sequence from action that occurred in the past, inserted in present continuity.

Flatbed editing machine A machine where the rolls of positive film, sound, music, and effects tracks are placed flat as on a table, rather than vertically as in a moviola.

Fliers Economical leaflets advertising a film.

Flips A new scene takes the place of the previous one as a card or page is flipped.

Freeze-frames The film continues to run with the last frame on screen as a still photograph.

Foley or effects track A sound track made up of all minor sounds occurring in the scene other than dialogue and music.

Four-walling Renting an individual theater for a short period to exhibit a film.

Frozen work print The definitive version of the images as they will be seen in the finished picture.

Fullcoat A 35mm tape used for sound recording where the entire width of the film is covered with magnetic material.

Gaffer A technician that handles all electrical and lighting for the cameraman.

Garden hose pan When the camera turns to right or left, then returns to former position.

Guide track A sound track taped on location for later use to synchronize sound in looping or dubbing session in the recording studio.

Head-on shot Actors approach the camera until they are in two-shot or close-up.

High angle The camera is placed above normal eye level to look downward.

Insert An extreme enlargement of a small part of a scene to convey information or maximum impact.

Intercuts Extended material inserted at appropriate time.

International track A mixed sound track with all the elements except dialogue.

Jump cut A cut or take out of chronological sequence that causes an unintentional break in continuity.

Letraset A patented form of adhesive letters of different sizes and styles.

Location list A list of locations mentioned in the screenplay, usually with a note after each one indicating where it will be filmed.

Long shot The camera sees the scene from afar, covering a large area.

Looping A system of recording dialogue on a special soundstage while viewing a projection of the scene.

Low angle The camera is placed below normal eye level to look upward.

Magnastripe A magnetic tape using regular 35mm perforated film base.

Majors The principal studios that produce high-budget pictures for mass distribution.

Master shot A single take that contains all of the complex movements of both actors and camera contained in a continuous action sequence.

Medium shot The camera sees the action as related to figures within the scene.

Montage A series of takes inserted in continuity to illustrate actions or establish passing of time.

Master shot a single take that contains all of the complex movements of both actors and camera contained in a continuous action sequence.

Minus dialogue track Same as the international track.

Negative cutter A specialist in cutting the original negative material to conform with the edited work print.

Negative insurance Insurance covering all costs of repeating a portion of a picture, the loss of which was caused by accident.

Neutral density filters Filters that reduce the amount of light entering the camera.

Odd or unusual angles The camera is usually not level in these shots.

Opticals Film effects obtained either in camera or with optical printers.

Outtake A take that is not used in the picture. It usually remains as a piece of negative. When necessary the editor may have it printed if the original take is not satisfactory.

Over-the-shoulder shots Shots that contain both the actor facing the camera and parts of the other actor in the shot.

Pan or panorama shot The camera turns to right or left to follow action or discover more visual territory.

Parallel cutting Cutting to action occurring at the same time at a different location.

Percentage deals When an actor accepts a percentage of the income of a picture in lieu of part or all of total salary.

Personal release All recognizable persons who appear in a film, other than those under specific contract to act in the picture, must sign a release conceding the right to use their appearance.

Pickup shot A shot unrelated to the presence of actors or crew, used to establish a location or action.

Point of view or pov. A shot where the action is seen from the position of one of the participants in the scene.

Polecat A tube that extends from floor to ceiling for mounting lights.

Pre-flashed (black box) Negative material that has been treated uniformly with a small quantity of light to modify its original characteristics, usually contrast.

Preliminary budget A rough estimate of the cost of the picture.

Press book An advertising pamphlet used to present the film mainly to distributors and exhibitors.

Principal photography All of the cinematography, excluding titles or pickup shots.

Prop or property list A list of all objects, furniture, firearms, cars, boats, planes, etc., used during the filming.

Property list Same as prop list.

Property release A document conceding the right to film in and/or around a recognizable real property such as a house, factory, or other local building.

Pulling focus The assistant cameraman changes the focus of the lens while the scene is being shot to maintain a clear view of the actor or the part of the scene so desired by the director.

Reel horse An apparatus to hold reels or pieces of film during the editing process.

Register pin movements A camera film movement that holds the film fixed solidly for a fraction of a second required to expose the frame.

Rerecording This is the process of mixing together all the separate sound tracks into one track.

Retake To film again a scene that was rejected because of a flaw.

Reverses Same as over-the-shoulder shots.

Rough cut Edited sequences with slates removed, but not cut to final form.

Rushes A positive print of scenes filmed on the day or days before projected completion, including frames with slate numbers.

Screen credits The relative positions occupied by the actors' names before or after the main title.

Screen tests Short scenes filmed of actors, to determine suitability for a role.

Script supervisor Formerly called script girl or continuity girl. A person in charge of maintaining continuity, avoiding discrepancies, and also representing a liaison between director, editor, and producer.

Selective focus Foreground in focus with background out of focus, or vice versa. Concentrates attention on only the part of the scene that is in focus.

Sequence A number of related scenes.

Setup Placing of camera, tracks, camera lights, etc., to film a specific shot within a sequence.

Shooting schedule The chronological sequence of shooting derived from and displaying information obtained from the breakdown.

Shooting script The screenplay, presented in the order in which it will be shot, without regard to final chronological order.

Slate A board with the name of the picture, director, and cameraman and specs for the number of the scene being filmed as well as the number of the take. It usually has a clapper, used to synchronize the sound.

Slug A piece of blank film inserted during editing in place of a shot not yet available.

Soundstage A large, soundproof building used for filming direct sound, as compared to a stage that in not soundproof.

Start date The date that an actor begins work on a picture.

Stills Photographs of artists or scenes from a picture, usually made on 8″ x 10″ black and white, glossy paper.

Stock shot A shot or sequence from pictures, newsreels, or documentaries, purchased for use in a new production.

Stop date The date that an actor finishes work on a picture.

Sub-distributors Distributors specializing in small geographical areas.

Subliminal cut Introducing one or more frames in a cut too short to be sensed by the human eye.

Subtitles Dialogue printed at the bottom of the frame in a language other than that spoken in the scene.

Suction mounting A device using a suction cup to fasten cameras to flat surfaces, such as car doors and hoods.

Synchronizer An apparatus with two to five sprocketed wheels, used to maintain the separate sound tracks in synchronism with the positive track during editing.

Tape splicer An apparatus to cut and join film together using transparent adhesive tape.

Tilt up The camera tilts up to view an area above the normal shot.

Tilt down The camera tilts down to inspect an area below the normal shot.

Timer A laboratory specialist who determines the density and the color filtering to be used in printing of the original negative to form the positive print that is projected in theaters.

Timing on the set The pace at which the actors and camera moves.

Timing in editing The pace at which cuts are made.

Titra A patented process where the emulsion is removed from the film to produce the subtitles.

Trailer A small reel consisting of specific parts of the film, to be used in theaters or television as advertising.

Two shots A shot containing two actors.

Typecasting Hiring an actor because his or her physical appearance coincides with a commonly accepted stereotype.

Walk-in shots The actors approach the camera.

Walk-away shots The actors walk away from the camera.

Wardrobe list A list of clothes, shoes, accessories, etc., needed for each character.

Wipes One scene pushes another from the scene.

Work print A print of the entire picture, used for editing purposes, usually in conjunction with several separate dialogue, sound effects, and music tracks.

Zoom shot A shot in which the size of the image is changed by a movement within the lens of the camera.

Index